Nanotechnology-Enhanced Orthopedic Materials

Related titles

Bone Repair Biomaterials
(ISBN 978-1-84569-385-5)

Joint Replacement Technology
(ISBN 978-0-85709-841-2)

Wear of Orthopaedic Implants and Artificial Joints
(ISBN 978-0-85709-128-4)

Woodhead Publishing Series
in Biomaterials: Number 102

Nanotechnology-Enhanced Orthopedic Materials

Fabrications, Applications and Future Trends

Lei Yang

ELSEVIER

AMSTERDAM • BOSTON • CAMBRIDGE • HEIDELBERG
LONDON • NEW YORK • OXFORD • PARIS • SAN DIEGO
SAN FRANCISCO • SINGAPORE • SYDNEY • TOKYO
Woodhead Publishing is an imprint of Elsevier

WP
WOODHEAD PUBLISHING

Woodhead Publishing is an imprint of Elsevier
80 High Street, Sawston, Cambridge, CB22 3HJ, UK
225 Wyman Street, Waltham, MA 02451, USA
Langford Lane, Kidlington, OX5 1GB, UK

British Library Cataloguing-in-Publication Data
A catalogue record for this book is available from the British Library

Library of Congress Control Number: 2015942344

ISBN: 978-0-85709-844-3 (print)
ISBN: 978-0-85709-850-4 (online)

For information on all Woodhead Publishing publications visit our website at http://store.elsevier.com/

Contents

Woodhead Publishing Series in Biomaterials

1 **Sterilisation of tissues using ionising radiations**
Edited by J. F. Kennedy, G. O. Phillips and P. A. Williams
2 **Surfaces and interfaces for biomaterials**
Edited by P. Vadgama
3 **Molecular interfacial phenomena of polymers and biopolymers**
Edited by C. Chen
4 **Biomaterials, artificial organs and tissue engineering**
Edited by L. Hench and J. Jones
5 **Medical modelling**
R. Bibb
6 **Artificial cells, cell engineering and therapy**
Edited by S. Prakash
7 **Biomedical polymers**
Edited by M. Jenkins
8 **Tissue engineering using ceramics and polymers**
Edited by A. R. Boccaccini and J. Gough
9 **Bioceramics and their clinical applications**
Edited by T. Kokubo
10 **Dental biomaterials**
Edited by R. V. Curtis and T. F. Watson
11 **Joint replacement technology**
Edited by P. A. Revell
12 **Natural-based polymers for biomedical applications**
Edited by R. L. Reiss et al
13 **Degradation rate of bioresorbable materials**
Edited by F. J. Buchanan
14 **Orthopaedic bone cements**
Edited by S. Deb
15 **Shape memory alloys for biomedical applications**
Edited by T. Yoneyama and S. Miyazaki
16 **Cellular response to biomaterials**
Edited by L. Di Silvio
17 **Biomaterials for treating skin loss**
Edited by D. P. Orgill and C. Blanco
18 **Biomaterials and tissue engineering in urology**
Edited by J. Denstedt and A. Atala
19 **Materials science for dentistry**
B. W. Darvell

20 **Bone repair biomaterials**
Edited by J. A. Planell, S. M. Best, D. Lacroix and A. Merolli
21 **Biomedical composites**
Edited by L. Ambrosio
22 **Drug–device combination products**
Edited by A. Lewis
23 **Biomaterials and regenerative medicine in ophthalmology**
Edited by T. V. Chirila
24 **Regenerative medicine and biomaterials for the repair of connective tissues**
Edited by C. Archer and J. Ralphs
25 **Metals for biomedical devices**
Edited by M. Niinomi
26 **Biointegration of medical implant materials: Science and design**
Edited by C. P. Sharma
27 **Biomaterials and devices for the circulatory system**
Edited by T. Gourlay and R. Black
28 **Surface modification of biomaterials: Methods analysis and applications**
Edited by R. Williams
29 **Biomaterials for artificial organs**
Edited by M. Lysaght and T. Webster
30 **Injectable biomaterials: Science and applications**
Edited by B. Vernon
31 **Biomedical hydrogels: Biochemistry, manufacture and medical applications**
Edited by S. Rimmer
32 **Preprosthetic and maxillofacial surgery: Biomaterials, bone grafting and tissue engineering**
Edited by J. Ferri and E. Hunziker
33 **Bioactive materials in medicine: Design and applications**
Edited by X. Zhao, J. M. Courtney and H. Qian
34 **Advanced wound repair therapies**
Edited by D. Farrar
35 **Electrospinning for tissue regeneration**
Edited by L. Bosworth and S. Downes
36 **Bioactive glasses: Materials, properties and applications**
Edited by H. O. Ylänen
37 **Coatings for biomedical applications**
Edited by M. Driver
38 **Progenitor and stem cell technologies and therapies**
Edited by A. Atala
39 **Biomaterials for spinal surgery**
Edited by L. Ambrosio and E. Tanner
40 **Minimized cardiopulmonary bypass techniques and technologies**
Edited by T. Gourlay and S. Gunaydin
41 **Wear of orthopaedic implants and artificial joints**
Edited by S. Affatato
42 **Biomaterials in plastic surgery: Breast implants**
Edited by W. Peters, H. Brandon, K. L. Jerina, C. Wolf and V. L. Young
43 **MEMS for biomedical applications**
Edited by S. Bhansali and A. Vasudev

44 **Durability and reliability of medical polymers**
Edited by M. Jenkins and A. Stamboulis
45 **Biosensors for medical applications**
Edited by S. Higson
46 **Sterilisation of biomaterials and medical devices**
Edited by S. Lerouge and A. Simmons
47 **The hip resurfacing handbook: A practical guide to the use and management of modern hip resurfacings**
Edited by K. De Smet, P. Campbell and C. Van Der Straeten
48 **Developments in tissue engineered and regenerative medicine products**
J. Basu and J. W. Ludlow
49 **Nanomedicine: Technologies and applications**
Edited by T. J. Webster
50 **Biocompatibility and performance of medical devices**
Edited by J-P. Boutrand
51 **Medical robotics: Minimally invasive surgery**
Edited by P. Gomes
52 **Implantable sensor systems for medical applications**
Edited by A. Inmann and D. Hodgins
53 **Non-metallic biomaterials for tooth repair and replacement**
Edited by P. Vallittu
54 **Joining and assembly of medical materials and devices**
Edited by Y. (Norman) Zhou and M. D. Breyen
55 **Diamond-based materials for biomedical applications**
Edited by R. Narayan
56 **Nanomaterials in tissue engineering: Fabrication and applications**
Edited by A. K. Gaharwar, S. Sant, M. J. Hancock and S. A. Hacking
57 **Biomimetic biomaterials: Structure and applications**
Edited by A. J. Ruys
58 **Standardisation in cell and tissue engineering: Methods and protocols**
Edited by V. Salih
59 **Inhaler devices: Fundamentals, design and drug delivery**
Edited by P. Prokopovich
60 **Bio-tribocorrosion in biomaterials and medical implants**
Edited by Y. Yan
61 **Microfluidic devices for biomedical applications**
Edited by X-J. James Li and Y. Zhou
62 **Decontamination in hospitals and healthcare**
Edited by J. T. Walker
63 **Biomedical imaging: Applications and advances**
Edited by P. Morris
64 **Characterization of biomaterials**
Edited by M. Jaffe, W. Hammond, P. Tolias and T. Arinzeh
65 **Biomaterials and medical tribology**
Edited by J. Paolo Davim
66 **Biomaterials for cancer therapeutics: Diagnosis, prevention and therapy**
Edited by K. Park
67 **New functional biomaterials for medicine and healthcare**
E. P. Ivanova, K. Bazaka and R. J. Crawford

68 **Porous silicon for biomedical applications**
Edited by H. A. Santos
69 **A practical approach to spinal trauma**
Edited by H. N. Bajaj and S. Katoch
70 **Rapid prototyping of biomaterials: Principles and applications**
Edited by R. Narayan
71 **Cardiac regeneration and repair Volume 1: Pathology and therapies**
Edited by R-K. Li and R. D. Weisel
72 **Cardiac regeneration and repair Volume 2: Biomaterials and tissue engineering**
Edited by R-K. Li and R. D. Weisel
73 **Semiconducting silicon nanowires for biomedical applications**
Edited by J. L. Coffer
74 **Silk biomaterials for tissue engineering and regenerative medicine**
Edited by S. Kundu
75 **Biomaterials for bone regeneration: Novel techniques and applications**
Edited by P. Dubruel and S. Van Vlierberghe
76 **Biomedical foams for tissue engineering applications**
Edited by P. Netti
77 **Precious metals for biomedical applications**
Edited by N. Baltzer and T. Copponnex
78 **Bone substitute biomaterials**
Edited by K. Mallick
79 **Regulatory affairs for biomaterials and medical devices**
Edited by S. F. Amato and R. Ezzell
80 **Joint replacement technology Second edition**
Edited by P. A. Revell
81 **Computational modelling of biomechanics and biotribology in the musculoskeletal system: Biomaterials and tissues**
Edited by Z. Jin
82 **Biophotonics for medical applications**
Edited by I. Meglinski
83 **Modelling degradation of bioresorbable polymeric medical devices**
Edited by J. Pan
84 **Perspectives in total hip arthroplasty: Advances in biomaterials and their tribological interactions**
S. Affatato
85 **Tissue engineering using ceramics and polymers Second edition**
Edited by A. R. Boccaccini and P. X. Ma
86 **Biomaterials and medical-device associated infections**
Edited by L. Barnes and I. R. Cooper
87 **Surgical techniques in total knee arthroplasty (TKA) and alternative procedures**
Edited by S. Affatato
88 **Lanthanide oxide nanoparticles for molecular imaging and therapeutics**
G. H. Lee
89 **Surface modification of magnesium and its alloys for biomedical applications Volume 1: Biological interactions, mechanical properties and testing**
Edited by T. S. N. Sankara Narayanan, I. S. Park and M. H. Lee

90 **Surface modification of magnesium and its alloys for biomedical applications Volume 2: Modification and coating techniques**
Edited by T. S. N. Sankara Narayanan, I. S. Park and M. H. Lee
91 **Medical modelling: the application of advanced design and rapid prototyping techniques in medicine Second Edition**
Edited by R. Bibb, D. Eggbeer and A. Paterson
92 **Switchable and responsive surfaces and materials for biomedical applications**
Edited by Z. Zhang
93 **Biomedical textiles for orthopaedic and surgical applications: fundamentals, applications and tissue engineering**
Edited by T. Blair
94 **Surface coating and modification of metallic biomaterials**
Edited by C. Wen
95 **Hydroxyapatite (HAP) for biomedical applications**
Edited by M. Mucalo
96 **Implantable neuroprostheses for restoring function**
Edited by K. Kilgore
97 **Shape memory polymers for biomedical applications**
Edited by L. Yahia
98 **Regenerative engineering of musculoskeletal tissues and interfaces**
Edited by S. P. Nukavarapu, J. W. Freeman and C. T. Laurencin
99 **Advances in cardiac imaging: techniques and applications**
Edited by K. Nieman, O. Gaemperli, P. Lancellotti and S. Plein
100 **Functional Marine Biomaterials: Properties and Applications**
Edited by Se-Kwon Kim
101 **Shoulder and elbow trauma and its complications: Volume 1: The Shoulder**
Edited by R. M. Greiwe
102 **Nanotechnology-Enhanced Orthopedic Materials: Fabrications, Applications and Future Trends**
L. Yang
103 **Medical devices: Regulations, standards and practices**
Seeram Ramakrishna, Lingling Tian, Charlene Wang, Susan Liao and Teo Wee Eong

Foreword

While today's conventional orthopedic implant devices have improved the quality of life for millions over the past several decades, it is becoming increasing clear that innovation is needed. In many cases, we are implanting the same orthopedic implant devices today that were implanted in the 1970s and earlier. However, an increasing number of people, a wider range of the population, larger age distributions, and even immune system compromised patients are now receiving orthopedic implants, whereas decades ago they were not. This has led to increased problems with orthopedic implants, including greater infection rates and persistent failure rates, which begs our attention for improvement.

This book beautifully highlights arguably the most innovative solution to come to orthopedics in a long time: nanotechnology. While the field of nanotechnology was first discussed in the early 1950s and signs of nanotechnology exist in Egyptian art, until now we have not fully appreciated the impact nanotechnology can have to increase bone growth, limit infection, and inhibit inflammation—all events that can increase orthopedic implant efficacy. Nanotechnology is the study of materials with fundamental length scales in the nanometer regime. By controlling materials at the nanoscale size, one can mimic the natural nanometer structures in bone and control surface energy to dictate cell functions.

Each chapter of this state-of-the-art book highlights the impact that nanotechnology has made and will continue to make in orthopedics, including how it has impacted almost every chemistry (from metals to ceramics to polymers and self-assembled materials) used as bone medical devices. Impressively, this book covers the most innovative nanomaterial fabrication techniques and emphasizes the safety of manufacturing and using nanomaterials in medicine—a problem that we do not yet fully understand and need to develop solutions for. It also provides a concise reason why we should consider using nanomaterials to regenerate bone in the first place, a rationale clearly progressing beyond the traditional trial and error mentality of conventional orthopedics.

This book will certainly be a must-have for all of those wishing to create novel solutions to our most persistent problems in orthopedics, allowing more patients to experience the benefits from bone medical devices.

Thomas J. Webster
President-elect, U.S. Society for Biomaterials
Mount Laurel
NJ, USA
The Arthur W. Zafiropoulo Chair and Professor
Department of Chemical Engineering
Northeastern University
Boston
MA, USA

Acknowledgments

I would like to thank my wife, Dr. Yanjie Bai, my parents and my sister, as well as other family members, for their incredible support during the writing of this book. I am also very grateful to Professors Thomas J. Webster, Brian W. Sheldon and David A. Stout for their constructive, effective, and supportive advice. I would also like to acknowledge the Jiangsu Provincial Special Program of Medical Science (BL2012004), the Priority Academic Program Development of Jiangsu Higher Education Institutions (PAPD), the National Basic Research Program of China (973 Program, 2014CB748600), the Jiangsu R&D Innovation Program (BY2014059-07), the National Natural Science Foundation of China (51472279), the Jiangsu Six Peak of Talents Program (2013-WSW-056), the Chinese Ministry of Education Start-up Fund for Overseas Scholars, the Hermann Foundation, and the National Science Foundation (award DMR-0805172) for supporting this work.

Fundamentals of nanotechnology and orthopedic materials

1

1.1 Introduction: nanotechnology and nanomaterials

Nanotechnology is the small science with big consequences. The prefix *nano* denotes a factor of one billionth (i.e., 10^{-9}). Therefore, nanotechnology is defined as the fabrication, manipulation, characterization, and application of materials or systems whose structures and components exhibit novel and significantly changed properties when control is gained at nanometer scale (specifically, <100 nm or $<10^{-7}$ m). Likewise, the material or matter that has a feature scale between 1 and 100 nm in at least one dimension is defined as nanomaterial. In fact, many natural matters or systems have hierarchical structures whose finest components are literally nanomaterials. One well-known example is natural bone, which comprises nanoscale calcium phosphate crystals and collagen fibrils in a highly ordered manner. Therefore, in a sense, nanotechnology is ubiquitous because nature creates, manipulates, and utilizes nanomaterials and nanoscale systems everywhere.

However, nanotechnology is still young in the history of human scientific research and engineering development. The brief history of the emergence and development of nanotechnology is summarized in Figure 1.1. Several years after the discovery of DNA helical structure that is also on nanoscale, Nobel laureate Richard Feynman purposely introduced the concept and idea of nanotechnology in the famous 1959 talk "There's Plenty of Room at the Bottom." Although the term *nanotechnology* was not mentioned in his talk, Feynman explicitly suggested that it would soon be possible to accurately manipulate atoms and molecules and even to create atomic-scale machines and factories [1]. Feynman's revolutionary vision was further elaborated by Taniguchi and Drexler in the 1970s, but the idea did not start turning into a new field of nanotechnology until the scanning tunneling microscope (STM) and the atomic force microscope (AFM) were invented in the 1980s.

With the assistance of the STM and AFM, a number of breakthroughs including the first manipulation of atoms at IBM and discoveries of iconic carbon nanomaterials (fullerene, carbon nanotubes, etc.) established a profound foundation for the rapid evolution and development of nanotechnology in the late 1980s [2]. After three decades of development, nanotechnology has become an enormously important interdisciplinary field encompassing physics, chemistry, engineering, materials science, biology, and medical science. It also becomes familiar to the public in the consumer markets of sports, electronics, and cosmetic products.

Interestingly, evolution and development of nanotechnology is closely bound to biology and medicine. In the Feynman's talk, he had already used DNA molecules to attest the possibility of storing a vast amount of information in an exceedingly

Nanotechnology-Enhanced Orthopedic Materials. http://dx.doi.org/10.1016/B978-0-85709-844-3.00001-X

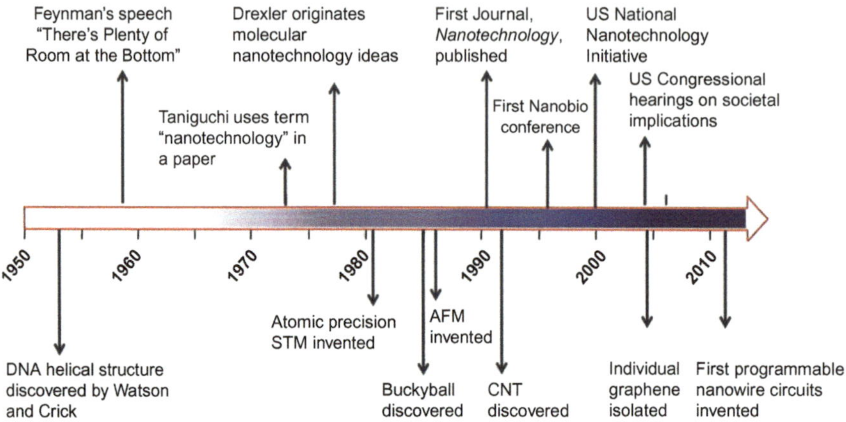

Figure 1.1 The brief history of nanotechnology. STM, scanning tunneling microscope; AFM, atomic force microscope; CNT, carbon nanotube [1].

small space, and also conceptualized a "mechanical surgeon" inside blood vessels for diagnosing and treating vascular diseases using the later-known nanotechnology. Since then, interactions between nanotechnology and life sciences have created countless opportunities for novel research tools, diagnostic methods, therapeutic strategies, and new molecules, materials, and systems. As a result, bionanotechnology and nanomedicine have become principal and fast-growing branches of nanotechnology within the past few decades.

This book will focus on a subfield of bionanotechnology and nanomedicine, nanotechnology-enhanced orthopedic materials, which is a specific interdisciplinary area of chemistry, physics, materials sciences, and orthopedics. This book will deliver a comprehensive overview of the field, infused with both great opportunities and challenges, with a particular interest in the fabrications, modifications, and applications of these materials. Furthermore, the book will explore novel nanomaterials with great potential for future applications, as well as the rationale and methodology of applying such materials for orthopedic purposes.

1.2 Fundamentals of fabrication and modification of nanomaterials

Nanotechnology-enhanced orthopedic materials are centered on a great number of nanomaterials and nanostructures with exceptional physicochemical, mechanical, and biological properties. Creation of appropriate nanomaterials is a key issue for their orthopedic applications. Herein, basic knowledge on general fabrication and modification of nanomaterials is presented. Starting in the next chapter, detailed fabrication and modification techniques of each category of nanomaterials for orthopedic uses will be introduced individually.

1.2.1 Fabrication strategies: top-down and bottom-up

Nanotechnologists generally use two opposite approaches for fabricating nanomaterials (also known as nanofabrication): top-down and bottom-up. These two strategies are not new and have already been used in manufacturing and engineering materials at larger scales (e.g., microelectronics). Nanotechnology has been arguably considered as a meeting point of two strategies, where both strategies reveal great effectiveness and necessity in fabricating wanted materials [3]. Figure 1.2 illustrates the top-down and bottom-up strategies and their scaling relationship to nanotechnology as well as examples of fabrication techniques.

1.2.1.1 Top-down strategy

Top-down strategy often refers to the manufacture of nanoscale materials or structures by machining, etching, milling, or successive cutting of large-sized bulk materials. There are plenty of widely used top-down approaches, which are generally divided into four categories: patterning, additive, subtractive, and comminution (break-down) techniques.

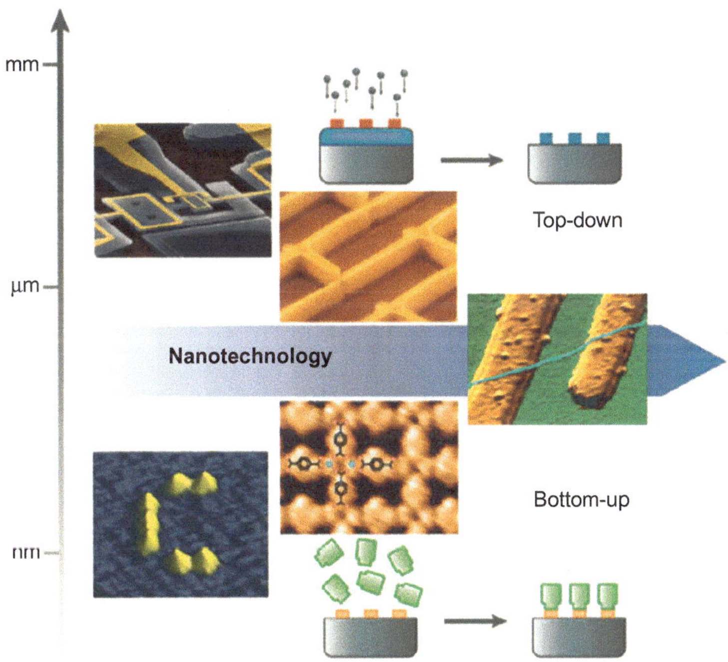

Figure 1.2 Top-down and bottom-up approaches for nanofabrication. Examples shown (clockwise from top) are an electron microscopy image of a nanomechanical electrometer obtained by electron-beam lithography, patterned films of carbon nanotubes obtained by microcontact printing and catalytic growth, a single carbon nanotube connecting two electrodes, a regular metal–organic nanoporous network integrating iron atoms and functional molecules, and seven carbon monoxide molecules forming the letter "C" positioned with the tip of a STM.
Reprinted from [4] with permission.

Patterning techniques

Frequently used patterning techniques include nanolithography, nanoimprint, and micro/nanoprinting. Nanolithography uses lights, charged ions, or electron beams to transfer a geometric pattern from a premade photomask to a photoresist layer, which is coated on a thin film material or the bulk of substrate. The technique then uses a series of post-treatments to chemically engrave the transferred pattern into, or allow the deposition of new material in the transferred pattern upon, the target material [5]. This process is illustrated in the schematic of the "Top-down" in Figure 1.2, where the photoresist and target material are marked in red and blue, respectively. What distinguishes between various nanolithography approaches is the type of radiation (photons, electrons, or ions) used to transfer the pattern. The radiation wavelength is a key to the spatial resolution of the approach. However, the final minimum feature size in the target material also depends on the quality and property of photomask and photoresist. Nanolithography techniques have demonstrated the capability to fabricate sub-10nm features [6].

In contrast to the nanolithography techniques that employ radiation to create patterns, nanoimprint uses a mechanical mold to delineate features, as illustrated in Figure 1.3. This technique generates nanoscale patterns by physically deforming a material and, therefore, can be used for the direct imprint of functional materials.

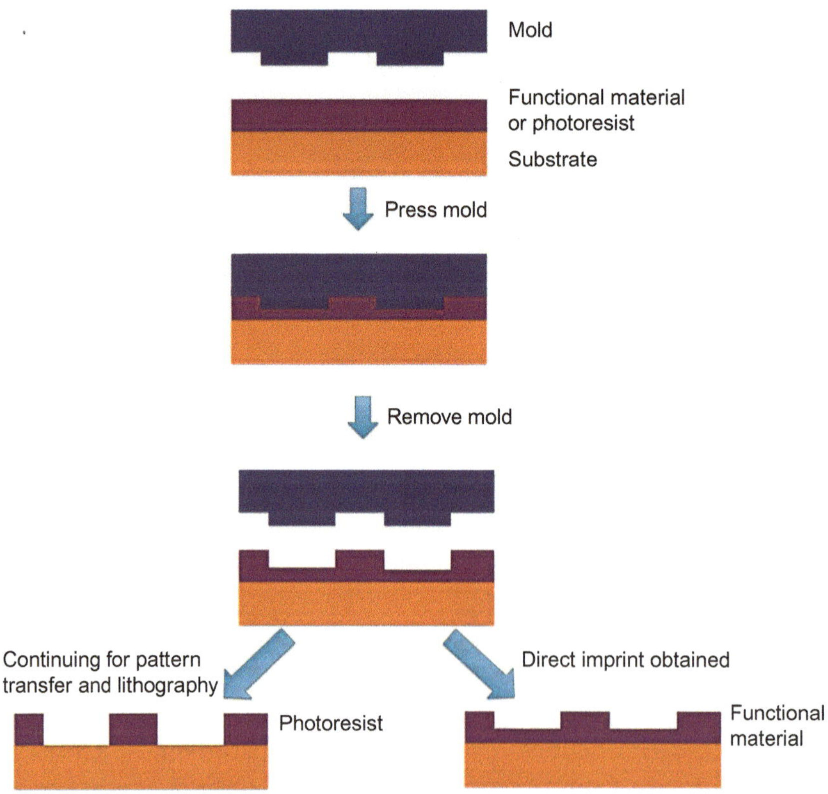

Figure 1.3 Illustration of nanoimprint techniques.

In a combined technique of nanoimprint and lithography, the material underwent imprinting can be photoresist and will be removed later as in conventional lithography (Figure 1.3). Nanoimprint has many advantages over radiation-based patterning techniques, including a high patterning resolution of sub-3 nm and capabilities of large-area patterning and 3-D patterning [6].

Micro- or nanoprinting (as known as soft lithography) uses a stamp with surface patterns to transfer a liquid ink onto a surface [7]. The feature resolution obtained from these printing techniques is poorer than that of nanoimprint, because the controls on ink flow and amount are difficult [6]. Improved nanoprinting technique uses solid material (solid ink) that is deposited on the stamp and the solid material pattern is separated from the stamp after bonding to the substrate [8]. Dip-pen nanolithography is another alternative to print nanomaterials. In this technique, an AFM tip is used as the "pen" to directly write molecules ("ink") on a substrate, where the molecules spontaneously transfer from the pen to the substrate by water meniscus [9].

Additive techniques

Although some patterning approaches (e.g., solid ink nanoprinting) can be also characterized as additive techniques, additive techniques mainly refer to deposition techniques. The common deposition techniques include physical vapor deposition (PVD), chemical vapor deposition (CVD), and atomic layer deposition (ALD), which typically require energy (thermal or electron/ion energies) input and vacuum environment. These techniques have been successfully used to fabricate nanoscale thin films, fibers, whiskers, and tubular or rod-like structures.

PVD is essentially a physical coating method that transports vaporized materials from a source onto a substrate in a vacuum chamber and the vapor condensates to form a thin film. Thickness of PVD thin film ranges from a few nanometers to even millimeters, usually depending on deposition time. PVD is also capable of fabricating multilayer thin films, gradient composition deposits, and freestanding structures [10]. Two representative PVD techniques are evaporation and sputtering, which mainly differ in the vaporization means. The evaporation relies on thermal energy (e.g., heating) while the sputtering uses gaseous ions (e.g., plasma) to vaporize source materials. Figure 1.4a shows a schematic of evaporation. These two techniques are widely accepted in nanomaterial fabrication due to high reproducibility and quality in products.

CVD, as indicated by its name, is a deposition method via chemical reaction of vapor phases. Depending on the means of inputting energy to deposition process, various CVD methods have been developed, such as thermal CVD, plasma-enhanced CVD, photo-assisted CVD, and low-pressure CVD. Figure 1.4b illustrates the example of thermal CVD system, in which reactive gas phases (usually referred to as precursors) flow into the vicinity of a substrate and react (or decompose) to form new materials on the substrate. The substrate can be pretreated or coated with catalysts and/or other reactive precursors to obtain desirable materials or structures. CVD techniques are versatile for fabricating nanomaterials of multiple dimensions (from 0-D to 3-D), highly ordered structures (from dots, wires, to scaffolds), and complex compositions (complex chemistries or compounds). These capabilities grant CVD immense potential in nanofabrication, especially in large-scale or large-quantity productions.

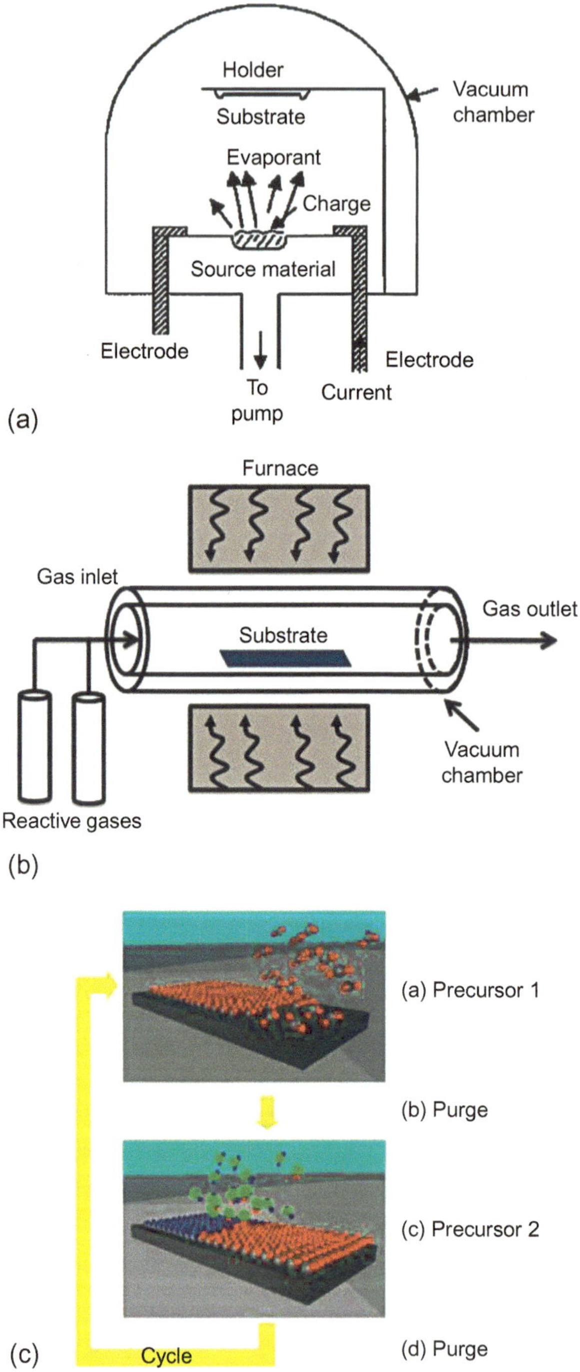

Figure 1.4 Schematics illustrate (a) evaporation, (b) CVD, and (c) ALD techniques. Figure (c) is reproduced from [11] with permission.

Table 1.1 **Examples of 0-D to 3-D nanomaterials fabricated by CVD approaches**

Dimension	Nanomaterial	Representative reference
0-D	Fullerene C_{60}	[12]
	Quantum dots (GaAs, InGaAs, Ge/Si, etc.)	[13–15]
1-D	Carbon nanotube (CNT)/fiber (CNF)	[16–18]
	ZnO nanorod	[19]
	Si nanowire	[20]
2-D	Nanocrystalline diamond (NCD)	[21]
	Graphene	[22]
	Diamond like carbon (DLC)	[23]
	Parylene thin film	[24]
3-D	DLC nanopillars	[25]
	CNT/CNF nanostructures	[26]
	ZnO nanostructures	[27]

Table 1.1 lists typical nanomaterials from 0-D to 3-D that can be produced by CVD techniques. In comparison to PVD, CVD techniques are more efficient for preparing nanocomposites and complex compounds (e.g., oxides, nitrides, etc.) [28]. More importantly, due to the nature of the gas process, the CVD technique can readily achieve uniform depositions or coatings on complex shapes and geometries such as curved surfaces, pore walls or blind holes, channels, and recesses [10]. This capability is extremely important for fabricating and modifying orthopedic implants and prostheses that usually have irregular shapes or architectures.

The ALD technique was first developed in the 1970s and has recently become an attractive fabrication approach for laminated nanomaterials. Essentially, the ALD technique employs sequential use of a gas-phase chemical process with an atomic-level precision [10], which is schematically shown in Figure 1.4c. In the figure, a thin film growth cycle for a binary compound from two gaseous precursors is illustrated [11]. Precise control of the deposited layer thickness renders ALD advantageous in fabricating novel nanostructures with extraordinary surface or interfacial properties. ALD is also attractive for fabricating biomaterials at low temperatures (<100 °C) or even at room temperature. For example, low-temperature ALD (LT-ALD) can be used to deposit coatings on temperature-sensitive polymers, which cannot readily be achieved by other deposition techniques. Recent studies have also demonstrated that LT-ALD enables the coating of biomolecules with desirable materials and the replication of biological nanostructures using synthetic materials [11,29]. These advances in the ALD technique open new opportunities for creating novel biomimetic materials and constructs.

Subtractive techniques

In contrast to the additive techniques, subtractive techniques remove regions of material from a coating on substrate or directly from a bulk material. Etching is one of the most frequently used subtractive techniques and it is also important in the lithography

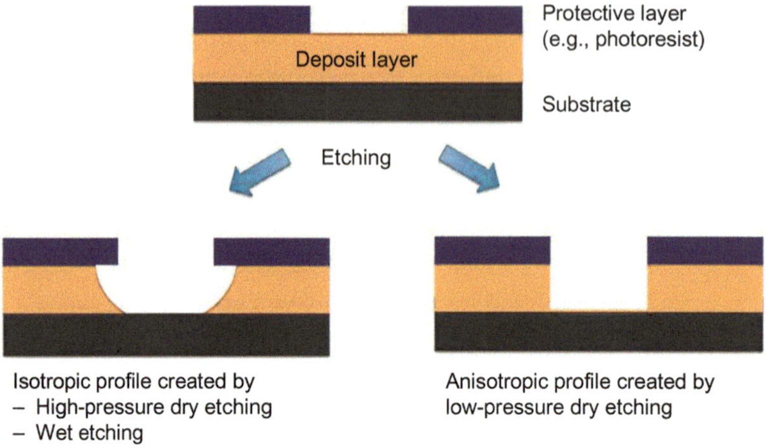

Figure 1.5 Isotropic and anisotropic etch profiles created by wet and dry etchings.

process in which etching removes the material not protected by photoresist after development or pattern transfer. Dry etching and wet etching techniques are two major types and differ in the form of etchants or chemicals used for removing materials [10]. In wet etching, a sample is immersed in liquid etchants or reactants and isotropic features whose profiles are just as wide as their depth are created. Dry etching, however, uses reactive gases or plasmas (e.g., reactive ion etching) to remove the materials chemically or physically and the resultant etch profile is determined by the etching pressure. For instance, isotropic and anisotropic etch profiles can be obtained by high-pressure and low-pressure plasma etchings, respectively [30]. Figure 1.5 shows the etch profiles created by the wet and dry etchings with different processing parameters.

Comminution (break-down) techniques

Comminution is the reduction of solid material particle size by fracture via grinding, milling, or similar processes. Comminution techniques are usually employed to produce particulate nanomaterials (e.g., powder) from larger-sized or bulk materials. Due to simplicity and low cost, attrition or grinding with the assistance of milling media such as milling balls has been used for producing nanoparticles or nanopowder since the late 1990s [31]. However, grinding techniques suffer several problems including particle agglomeration, large residual internal stress, surface defects, and contaminations in the final products.

To overcome these problems, another comminution technique—atomization—was developed. Atomization produces solid or aerosol particles with reduced sizes by spraying molten material or material solution or suspension under conditions such that it breaks down and then solidifies as fine powder [32]. In a typical atomization process, a molten material passing through a nozzle scatters into fine droplets by a high-speed medium (e.g., gas or water) and then the droplets solidify to powder. Obviously, the atomization technique is highly efficient for preparing micron and submicron powders at industrial scales and recent development has enabled atomization to produce nanoparticles of sizes down to 20 nm [33].

1.2.1.2 Bottom-up strategy

When material feature size goes to the sub-10 nm range, top-down fabrication techniques encounter difficulties essentially because the fabrication tools are too large to deal with such tiny objects. As feature size decreases, bottom-up strategy becomes more convenient and efficient to prepare nanomaterials compared to the top-down strategy. The bottom-up strategy builds up a material from smaller scales and often refers to self-organization and/or self-assembly of atomic or molecular components into nanoscale or larger structures (lower part of Figure 1.2). The bottom-up approaches can produce nanostructures of a size range from a few nanometers or less up to thousands of nanometers. As a matter of fact, the bottom-up strategy has long been employed by biological systems where the nature harnesses the chemical and physical forces to create the structures of life [34]. As nanotechnology emerges, a large number of man-controlled bottom-up approaches have been developed for fabricating nanoparticles, nanowires, and nanoassemblies. Some of the representative bottom-up approaches are introduced as follows.

Colloidal synthesis

Colloid is a homogeneous system consisting of nanoscale (1–100 nm) or mesoscale (from 100 nm to hundreds of microns) molecules or particles dispersed through a medium (or second phase). In this book, colloidal synthesis is a general term describing the chemical synthesis route for preparing nanomaterials of various dimensions and geometries through a colloidal process [30]. Figure 1.6 illustrates the colloidal synthesis route of Au nanoparticles with two geometries, in which Au^{3+} ions are reduced to Au atoms and their aggregates nucleate and continuously grow into different geometries in the presence of different surfactants. Such nanoparticles of adjustable sizes, shapes, and geometries can be further used as templates to fabricate other nanostructures [34], which will be discussed later.

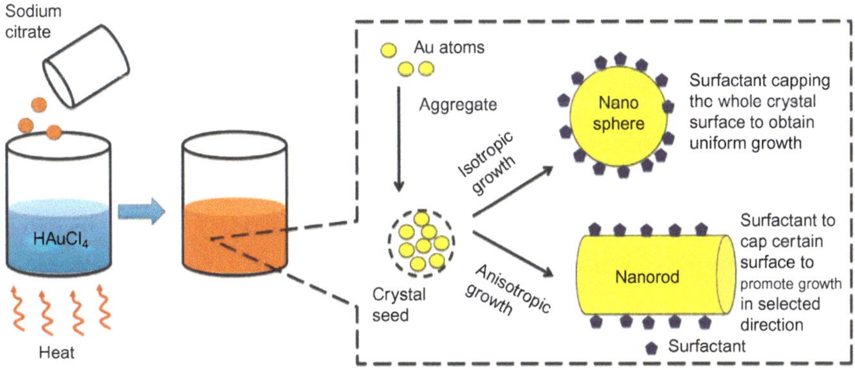

Figure 1.6 Schematic of colloidal syntheses of Au nanospheres and nanorods. Different surfactants are used to cap surfaces (or crystallographic facets) of crystal seed, resulting in altered growth mechanisms and leading to two geometries of Au nanoparticle.

Unguided self-assembly

Self-assembly is a spontaneous process in which materials form structures themselves as a result of the local energy minimum of the material system [6]. More details about self-assembled nanostructures will be covered in Chapter 6. Unguided self-assembly does not use guiding templates and usually results in a structure whose domains (sub-area of a large material or system) are randomly oriented but components within the domains are uniformly arranged. A representative of unguided self-assembled material is micelle. Micelle is a colloidal aggregate of a certain number of molecules that have opposite polarity or wettability (i.e., hydrophobicity) at two ends. Such molecules can assemble into various shapes or structures, as shown in Figure 1.7. The shape and size of a micelle are functions of its component molecule geometry and solution conditions, such as molecular concentration, temperature, and pH. By loading chemical or biological entities inside, micelles are widely used for delivery and controlled releases of drug, gene and cell [35].

Template-assisted self-assembly

Unguided self-assembly has little control on domain orientation, rendering the fabrication of ordered nanostructure difficult. To overcome this problem, templates can be utilized to guide self-assembly and to obtain better controls of regularity or orders of resultant nanostructures. The template can be either smaller or larger than the final self-assembly. For patterning purposes, the template (i.e., substrate with guidance) is

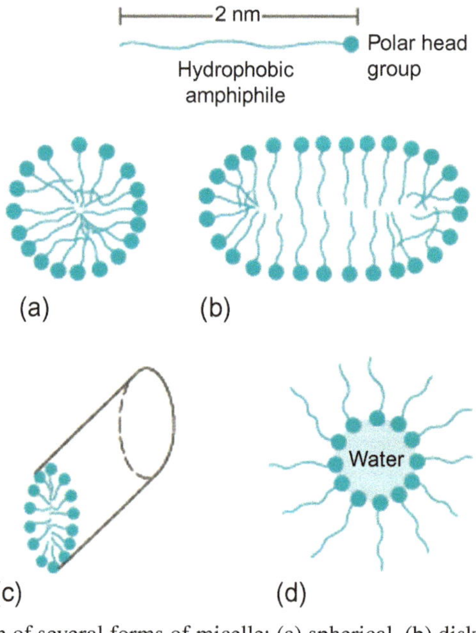

Figure 1.7 Illustration of several forms of micelle: (a) spherical, (b) disk, (c) rod, and (d) inverse (or reverse).

often larger than the final assembly and the patterning or guiding cues are prefabricated on the template. In this manner, the self-assembly can precisely follow the guidance and establish a desirable orientation within the designated area of the substrate. For instance, self-assembled monolayer (SAM) technique is a process in which only one layer of molecules is attached to a substrate. This technique uses molecules with functional groups at two ends where one group attaches to the substrate while the other does not [6]. Therefore, the size, pattern, and coverage of self-assembly on the substrate can be tailored by altering the guiding patterns and other guiding cues on the substrate.

SAM is also capable to coat nanoparticles and in this case the nanoparticles act like templates whose sizes are comparable to the final self-assembled structures. Biomolecules can also be used as templates for self-assembly. For instance, DNA-templated assembly has been developed to arrange material building blocks through base-pair recognition [36,37]. Figure 1.8 illustrates the two approaches of DNA-templated assembly: (a) material components made with complimentary DNA strands assemble together only when the right pairing is present [38], and (b) complimentary DNA strands assemble into supermolecular structures first (nanogrid in Figure 1.8b1) and functional material components incorporate later via self-assembly.

Phase separation approaches

Another group of bottom-up approaches relies on simultaneous phase separation caused by difference in material miscibility. Block copolymers revealing varied

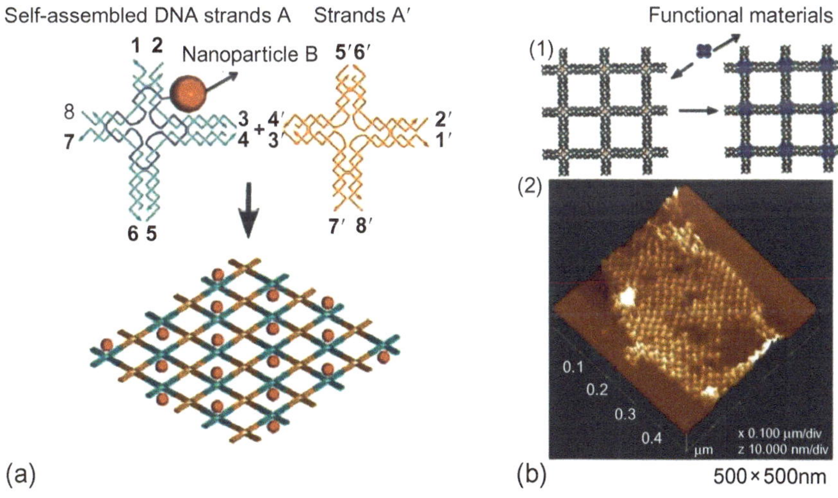

Figure 1.8 Two approaches of DNA-templated self-assembly: (a) self-assembled DNA strands A (blue cross shape) carrying nanoparticle B pair with another self-assembled DNA strands A′ to form a 2-D nanogrid with nanoparticles (adapted from [38] with permission); (b) Complimentary DNA strands pair into nanogrids and functional materials incorporate with the nanogrid later (b1), forming arrays of nanomaterials (b2) (adapted from [37] with permission).

miscibilities around a certain temperature (e.g., order–disorder temperature) can be used for preparing nanostructures via the phase separation process. In this process, different copolymer blocks are first connected together and mixed well with each other under a certain temperature. When the copolymers are heated above that temperature, phase separation occurs and the different block copolymers become nonmixable; however, because the block copolymers have been connected at their ends, internal stress evolves and forces the spatial dislocation of polymer phases, resulting in ordered or periodic patterns. This approach can also be combined with other bottom-up approaches mentioned above to create complex structures. For instance, Figure 1.9a shows uniformly distributed nanoparticles assemble into strips with various widths as a result of phase separation of block copolymer on a substrate with guiding strip patterns [39,40].

Phase separation and transfer strategies have also been developed for the synthesis of inorganic and polymeric nanocrystals. For example, the strategy of liquid–solid-solution (LSS) phase separation and transfer has demonstrated exceptional capacities to assemble a wide range of materials with novel structures and functions [41]. As demonstrated in Figure 1.9b, the LSS process involves the reduction of noble ions at the interfaces of liquid, solid, and solution phases at designated temperatures. This LSS phase transfer and separation process can generate nanocrystals with a variety of chemistries and properties, such as metal, semiconducting, fluorescent, magnetic and dielectric, and conducting polymer nanoparticles.

It is worth noting that the boundaries between numerous bottom-up approaches are sometimes ambiguous, so combining multiple approaches to obtain desirable or complex nanostructures is often necessary. For example, the colloidal process is often combined with template-assisted self-assembly to produce hollow nanospheres and other complex structures [42,43]. This versatility of bottom-up approaches opens many opportunities to fabricate a myriad of nanomaterials with extraordinary properties.

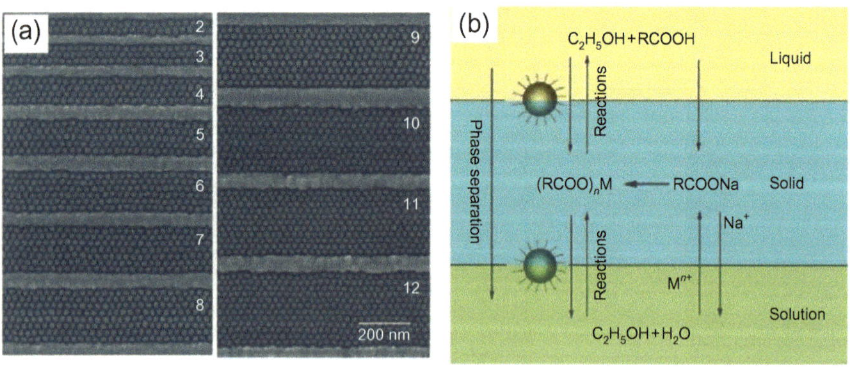

Figure 1.9 (a) Phase separation of diblock copolymer (polystyrene-β-polyferrocenyldimethylsilane) guided by substrate strips (adapted from [39] with permission); (b) scheme of liquid–solid-solution (LSS) phase separation and transfer synthetic strategy (reproduced from [41] with permission).

1.2.1.3 Combination of top-down and bottom-up strategies

Both top-down and bottom-up strategies have their own advantages and disadvantages, but these two strategies can be complementary. Top-down approaches have better controls of process and are facile for industrial manufacture of nanomaterials, but they are likely to introduce internal stress, surface defects, and contaminations in final products. On the contrary, bottom-up approaches promise a better chance to obtain nanostructures with less defects, more homogeneous chemical composition, and atomic-level precision, although the mesoscopic or macroscopic controls of assembly geometry and process are usually difficult. Therefore, the combination of both strategies can overcome some of their shortcomings and provide rich possibilities for innovative nanofabrication methods. For instance, the template-assisted self-assembly is combined with top-down nanolithography to obtain the templates with precise spatial control, allowing the creation of complex hierarchical structures and devices [44].

1.2.2 Modification of nanomaterials

Modification of nanomaterials is one of the most important research areas in nanotechnology because it opens lots of avenues to impart new functionalities and properties to nanomaterials. In general, there are two types of modification: bulk modification and surface modification. Bulk modification usually refers to that in the chemistry, constituent, structure of the bulk of material, which is more often associated with fabrication processes of nanomaterials. Surface modification, however, is limited to the surface region of the material where the resultant properties after modification may differentiate from the bulk properties.

The interface between biomaterials and biological systems (e.g., cell, tissue, etc.) is a thin transitional layer where changes in material surface properties can greatly affect cellular or tissue responses and functions. Therefore, surface modification can equip a material with desirable biological properties while retaining its bulk properties (e.g., mechanical properties). This is why surface modification has now received much more attention than bulk modification in the field of orthopedic materials. This section focuses on some fundamentals and important surface modification techniques, including coating deposition, implantation and bombardment, molecular grafting, and enzymatic modification.

1.2.2.1 Coating deposition

Functional coating is an external layer of materials deposited on the surface of a bulk material to provide additional functionality. For some orthopedic implants, coatings become a key for enhancing biological and tribological performance of the implant. For instance, wear-resistant coating is fabricated at the articulating surface of the implant to reduce abrasion and wear, while bioactive calcium phosphate coating is devised to interface with bone to achieve better osseointegration. Common coating deposition methods include vapor deposition and wet-chemical deposition techniques. PVD and CVD are frequently used vapor deposition techniques (see Section 1.2.1.1), while sol–gel and precipitation techniques are representative wet-chemical deposition methods.

Sol–gel techniques can produce nanocrystalline coatings with high purity, chemical homogeneity, and nanometer-range thickness due to the molecular mixing nature of the process. In sol–gel process, colloidal sol of precursors (e.g., metal alkoxides and metal chlorides) evolves gradually to form a gel-like network, and the gel-like network can be readily coated on a surface and then transferred into a uniform coating via heat treatment [45]. The sol–gel techniques are versatile for tailoring surface properties and combining new functionalities into a material. There is a long list of coatings that can be prepared by sol–gel techniques, from oxides (Al_2O_3, SiO_2, TiO_2, ZrO, ZnO, etc.) to apatites (hydroxyapatite, fluorhydroxyapatite, and fluorapatite, etc.) and to nitrides (Si_3N_4, AlN, etc.) [45–47]. In addition, sol–gel techniques have relatively simple processes (such as dip coating, spin coating, and padding), which can be readily added and piggybacked to other fabrication techniques.

Precipitation techniques are suitable for preparing large areas of biocompatible or bioactive coatings at low temperatures (e.g., body temperature). The techniques usually include aqueous reactions of different reagents that form solid-state precipitates and *in situ* deposition of the precipitates onto a material. For example, hydroxyapatite nanocrystals can be produced by the coprecipitation of $Ca(NO_3)_2$ and $(NH_4)_2HPO_4$ mixture and further deposited onto titanium implant surface to form an osteoconductive nanocoating [48].

1.2.2.2 Discharge and plasma treatments

Discharge and plasma treatments disassociate gas molecules to generate reactive species (e.g., radicals, ions, photons) and terminate the reactive species onto material surfaces to achieve specific functionalities. A typical setup of discharge or plasma treatment is shown in Figure 1.10. Gas species employed in the discharge and plasma treatments include oxygen, hydrogen, nitrogen, ammonia, argon, and reactive monomers [45]. The nanomaterials that are commonly modified by discharge

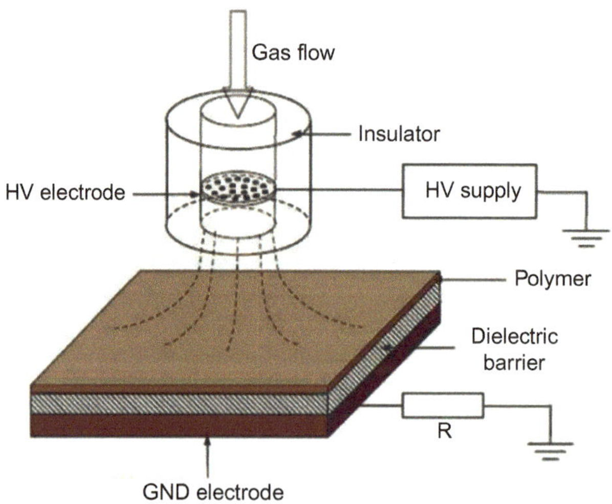

Figure 1.10 Typical setup for discharge/plasma treatment of a polymeric surface. Reprinted from [49] with permission.

or plasma treatments range from oxides to polymers to carbon nanostructures (e.g., carbon nanotube, carbon nanofiber, nano diamond, graphene, etc.) [50–53]. Discharge and plasma treatments are particularly effective on modifying nanomaterial wettability and terminating nanomaterial surfaces with radicals such as –OH, –H, and $-NH_2$ that are pertinent to biological responses [50]. In addition, plasma and discharge surface treatments have also been used to improve tribological property like wear resistance of the metallic components in orthopedic implants [54].

1.2.2.3 Molecular grafting

In order to enhance biocompatibility properties of implant materials and promote tissue growths, functional proteins, peptides, and other bioactive molecules are frequently grafted onto material surfaces. This strategy is often known as functionalization and has been extended to modify nanomaterials with the assistance of numerous grafting techniques. The plasma treatment introduced above can be adapted to graft biomolecules or polymers onto nanotubes and nanoparticles [55,56]. Other molecular grafting methods are based on chemical reactions using coupling reagents or by covalent binding. Researchers have demonstrated that a wide spectrum of chemistries are eligible for grafting purposes, including poly(ethylene glycol) (PEG), antibodies, DNA molecules, hydroxyl containing bipyridine molecules, and others [45,57,58]. PEG has been extensively used as a grafting reagent for fastening proteins, peptides, and functional polymers upon material surfaces, since it is nontoxic, nonimmunogenic, nonantigenic, highly soluble in water, and resistant to renal clearance. The grafting process using PEG is also known as PEGylation, and numerous PEG derivatives offer rich opportunities to conjugate a variety of biomolecules to nanomaterial surfaces [59].

1.2.2.4 Enzymatic modification

Recently, a new approach using enzymes to modify material (or nanomaterial) surfaces has been developed. This approach utilizes enzymatic hydrolysis or degradation to mildly "etch" the material surfaces and thus create desirable surface structures or properties [45]. Many polymers, including polyvinyl chloride (PVC), polyacrylonitrile, polyethylene terephthalate, and polyamides, have been modified by this method. In a recent study, PVC tubes were soaked in Rhizopus arrhizus lipase to create nanometer surface features that exhibited antibacterial properties [60]. Enzymatic modification demonstrates great potential and promise for biomedical and nanomedicine applications due to its biocompatible, environmentally friendly, and simple process.

1.3 Interactions between musculoskeletal tissue and biomaterial

In the musculoskeletal system, bone is the primary tissue/organ interacting with prosthetic implants/biomaterials and their interface is a crucial region where the interactions pertinent to new tissue formation and implant efficacy occur. Bone is a complex biological system that comprises both hierarchical structures and living bone-remodeling components. The architecture of bone is composed of nanoscale fibrous

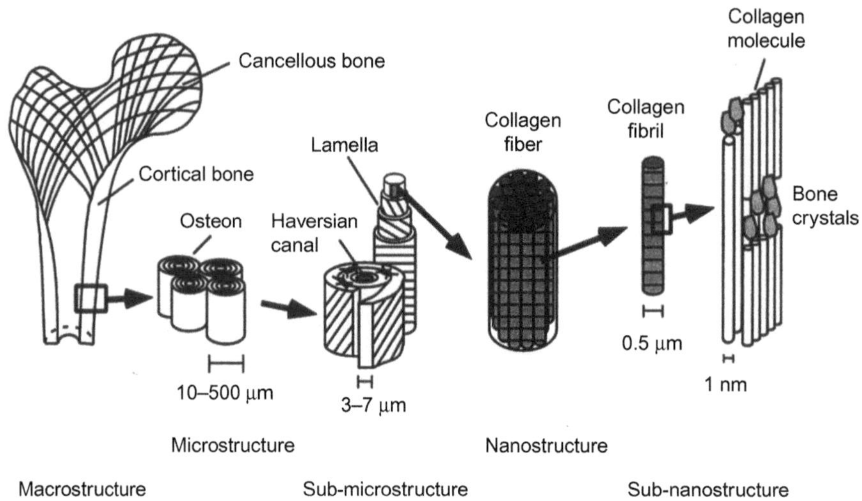

Figure 1.11 Hierarchical micron-to-nanostructure of human femur.
Adapted from [61] with permission.

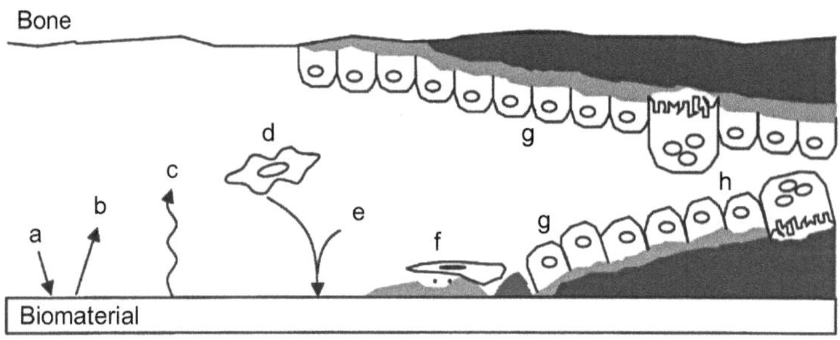

Figure 1.12 Schematics illustrating interfacial interactions between bone tissue or cell and implanted biomaterials viewed from the host tissue perspective: (a) protein adsorption from blood and tissue fluids, (b) protein desorption, (c) substrate surface changes and material release, (d) inflammatory and connective tissue cells approach the implant, (e) possible targeted release of matrix proteins and selected adsorption of proteins, (f) formation of lamina limitans and adhesion of osteogenic cells, (g) bone deposition on both the exposed bone and implant surfaces, and (h) remodeling of newly formed bone.
Reprinted from [62] with permission.

collagen matrices, noncollagenous proteins, and nanocrystals of calcium phosphates (mainly hydroxyapatite). Figure 1.11 shows the hierarchical micron-to-nanostructure of natural bone. Cellular components of bone include three types of cell: osteoblasts (bone-forming cell), osteocytes (bone-maintaining cells), and osteoclasts (bone-resorbing cells).

Basically, all of the events occurring at the tissue–material interface can be categorized as biological responses toward the implant material (host tissue responses) and, conversely, material responses to the host tissue [62]. These two categories can be further viewed from the host tissue and the biomaterial perspectives as illustrated in Figures 1.12 and 1.13, respectively.

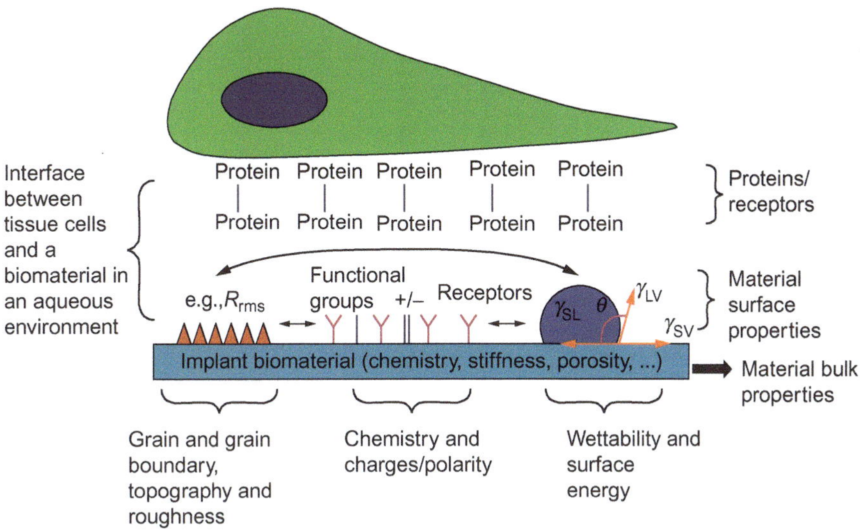

Figure 1.13 Schematic illustrating potential interfacial interactions between a cell (or tissue) and implanted biomaterials viewed from the material perspective. Double arrows indicate interactions among surface properties of biomaterials. In the schematic, examples of nanomaterial surface properties include root mean square roughness (R_{rms}), surface electrical charges (+/−), contact angles (θ), interfacial tension between solids and vapors (γ_{SV}), interfacial tension between solids and liquids (γ_{SL}), and interfacial tension between liquids and vapors (γ_{LV}).

1.3.1 Biological responses to biomaterials

As illustrated in Figure 1.12, biological responses toward implant materials (including nanomaterials) comprise a series of complicated physiological activities triggered at the biomaterial–tissue interface at a time scale from milliseconds (e.g., protein adsorption and desorption) to even months (e.g., foreign body reaction and fibrous encapsulation). For orthopedic implants, there are several important host responses occurring after implantation. These responses include inflammatory reactions, immunological responses, toxicity, blood–surface interactions, thrombosis and infection, which will be extensively discussed in Chapter 9. To a great extent, these host responses include aberrations of common host defense processes [63]. Several host responses pertinent to orthopedic implant applications are briefly introduced below.

Inflammatory responses occur immediately after the injury of tissues or organs, including a sequence of events such as acute and chronic inflammation, granulation tissue generation, foreign body reaction, and fibrous encapsulation (or fibrosis). The progress of inflammation is mediated by a number of chemical mediators released from plasma, cells, and tissues. These chemical mediators also play an important role in the degradation of implant materials, and in turn, degradation products may initiate increased release of chemical mediators to prolong the inflammatory response, which ultimately can cause implant failure. Besides inflammation, implant materials (especially smaller-sized materials like nanomaterials) may trigger responses from the

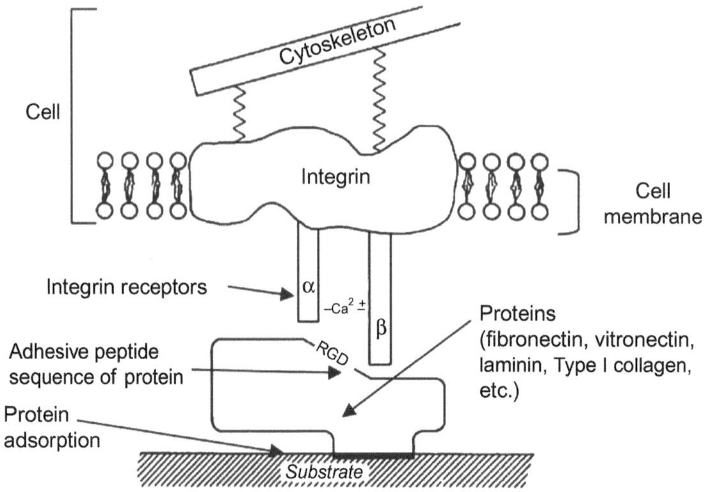

Figure 1.14 Schematic representation of protein-mediated cell adhesion on biomaterial surfaces. Adapted from [65].

immune system, which recognizes and differentiates these "foreigners" from the host, and protects the host from exposure to them by destroying these foreign substances.

Due to contaminations during the operation process, infection may also initiate before the bonding between bone and implant occurs. In many cases, there are multiple bacteria leading to the infection of implants. For example, *Staphylococcus epidermidis* exists on the human skin and can infiltrate to implant sites to cause infection. In contrast, *Staphylococcus aureus* is a natural tissue pathogen that may cause infection due to tissue damage around implants.

It is worth emphasizing that a great number of proteins (adhesive proteins, cytokines, growth factors, etc.) are involved in the interactions between materials and biological systems in an aqueous environment (bodily fluids, culture media, etc.). The role of interfacial proteins in mediating cell and tissue responses to biomaterials is exceptionally important [64]. Specifically, before cells adhere to a material surface, proteins will adsorb onto the surface within milliseconds to potentially interact with select cell membrane receptors, as shown in Figure 1.14 [65]. Therefore, the accessibility of cell adhesive domains (such as specific amino acid sequences of adsorbed vitronectin, fibronectin, and laminin) may either enhance or inhibit subsequent cellular adhesion and growth on a material. Thus, selective protein adsorption and its accessibility to cells are key factors toward understanding the biological responses to biomaterials. More details on the role of proteins in biological responses of nanomaterial will be analyzed in Chapter 8.

1.3.2 Nanomaterial properties affecting biological responses

From materials science perspective, biological responses of tissues or cells are strongly correlated to material properties. Figure 1.13 illustrates the interface between a cell and an implant material, showing the possible interactions between the cell membrane, proteins, and material properties. In this section, some important properties of nanomaterials that are determinant of cell or tissue responses are introduced.

1.3.2.1 Size and surface area

Compared to conventional micron-scale materials, extremely small sizes of surface feature or nanostructured component in materials can lead to numerous distinct biological responses. The small grain sizes of material greatly affect its surface topography and roughness, which is discussed later. More importantly, greatly increased surface area of nanomaterials as a result of extremely small grain sizes (or feature sizes) can significantly increase material surface reactivity or surface affinity for the adsorption of proteins or biological entities, leading to enhanced or mediated adhesion of cells or bacteria and promoted catalytic biochemical reactions (e.g., enzymatic reactions during inflammation and immune responses).

1.3.2.2 Topography and roughness

Surface topography is a predominant material property that has been shown to alter material biological responses. Although material topography is difficult to define by a simple parameter, roughness is a widely used parameter whose effects on tissue/cell responses have been studied [66]. An important lesson learnt from natural biological systems is that many tissues (e.g., bone, tooth, and blood vessels) have inherent nanoscale topographies and roughness due to their hierarchical micron-to-nanostructures. Based on this information, biologically inspired nanostructured materials have been shown to enhance tissue and bone cell responses and promote the efficacy of prosthetic implants. For example, carbon fibers, alumina, and poly(L-lactic acid) with nanoscale surface roughness (e.g., root mean square roughness values from 0.5 to tens of nanometers) have increased osteoblast adhesion, proliferation, alkaline phosphatase activity, and calcium deposition compared to the same chemistry of larger scales [67–69].

However, the mechanisms behind nanotopographically mediated tissue responses are still not clear. The increased adsorption of cell adhesive proteins (such as fibronectin, vitronectin, etc.) on nanoscale rough surfaces, due to either uneven surface landscapes or increased specific surface area, is a plausible mechanism. However, this mechanism is not enough to explain all the nanoscale roughness (or nanotopography) related tissue responses. For example, there is a lack of evidence to explain the fact the many different types of cells "see" the same roughness but their responses are different from cell type to cell type [64]. These questions will be further examined in Chapter 8.

1.3.2.3 Surface chemistry

In addition to nanoscale topography and roughness, biological responses are very sensitive to the changes in biomaterial surface chemistry. For example, Garcia et al. [70] reported that OH- and NH_2-terminated surfaces (SAMs of alkanethiols on gold) up-regulated osteoblast-specific gene expression, alkaline phosphatase enzymatic activity, and matrix mineralization compared with surfaces presenting COOH and CH_3

groups. The inherent (or bulk) chemistry of materials usually refers to their chemical or phase composition and crystallinity, while the surface chemistry often refers to grain/grain boundary chemistry and physical or chemical bonds formed on the surface. In many cases, the surface chemistry is not necessarily the same as the bulk chemistry, which has been mentioned in the previous section of surface modification and will be elucidated in the following chapters.

1.3.2.4 Surface wettability and surface energy

It has long been observed that cell adhesion and subsequent activities reveal dependence on implant surface wettability [71]. Surface wettability (i.e., the extent of the surface hydrophilicity or hydrophobicity) has been traditionally determined by water contact angles (θ) as Vogler suggested a definition of biologically hydrophilic surfaces having a θ less than 65° [72]. θ is closely related to surface energy and a classic definition of θ is given by the Young's equation (as illustrated in Figure 1.13):

$$\cos\theta = (\gamma_{SV} - \gamma_{SL}) / \gamma_{LV} \tag{1.1}$$

where γ is surface energy and γ_{SV} is the surface energy (or surface tension, in the unit of J/m^2) between solid and vapor, γ_{SL} is the surface energy between solid and liquid, and γ_{LV} is the surface energy between liquid and vapor. This equation indicates that the ways to change surface wettability essentially include manipulating the three surface energy terms mentioned above. Modification of nanomaterial surface chemistry is a common approach used to alter nanomaterial surface energy and resulting wettability. Moreover, adjustment in nanometer surface topography is also a possible means to alter nanomaterial surface energy and wettability [73].

Many studies so far have demonstrated a proportional relationship between surface energy of nanomaterials and the adsorption of hydrophilic cell adhesive proteins (e.g., fibronectin and vitronectin). For example, maximum vitronectin adsorption was noted on hydrophilic surfaces compared to hydrophobic ones [65]. Therefore, it is speculated that hydrophilic (i.e., high surface energy) nanomaterials have a higher affinity for cell adhesive proteins and, subsequently, promote cell functions and tissue responses better than hydrophobic nanomaterials. The verifications and debates on this hypothesis will be further discussed in the following chapters.

1.3.2.5 Other nanomaterial properties

Mechanical properties of materials have recently been correlated to cellular or tissue responses [74,75]. To date, studies have demonstrated that stem cells and tissue cells (such as skin, muscle, and brain cells) sense and respond to local matrix (e.g., extracellular matrix or synthetic material) stiffness through the formation of molecular adhesion complexes and changes in the actin–myosin cytoskeleton, which provide a feedback of the matrix stiffness for cell adhesion, motility, and differentiation [74]. For example, epithelial cells and fibroblasts on a collagen-coated polyacrylamide substrate that allows the stiffness to be altered (elastic moduli from ~5 to ~80 Pa) revealed less cytoskeletal spreading and higher rates of motility or lamellipodial activity on flexible substrates than rigid substrates [76]. The study also showed that focal adhesion of cells on flexible substrates was irregularly shaped and highly dynamic, whereas those on stiff substrates had a normal morphology and were much more stable.

Electrostatic interaction also plays a role in mediating biological responses to biomaterials because cell membrane is usually charged and virtually all material surfaces are charged in aqueous solutions. Surface charges on nanomaterials can be induced or modified by altering surface chemistry or charging the nanomaterials in an electrical field. However, the effects of surface charge on cell functions are not clear [64]. For example, a study revealed that positively charged indium tin oxide enhanced the adhesion of rat marrow stromal cells but impaired subsequent cell spreading and differentiation [77]. In addition, there is some early evidence showing the influence of piezoelectric and/or conductive materials on osteoblast responses. For instance, Itoh et al. [78] created electrically polarized hydroxyapatite with pores to increase bone growth and decrease osteoclast activity, and they attributed these effects to the electrical polarity of material surfaces with pores.

1.4 Summary

The emergence of nanotechnology and nanomaterials in the past few decades has impacted and reshaped many aspects of orthopedic practice and research. Nanotechnology has demonstrated exceptional promise for orthopedic medical applications due to its unique capability to devise and fabricate materials and devices with extraordinary properties. However, problems and challenges remain for nanotechnology before it achieves wide and extensive application in clinical practices. The rest of this book will focus on both the opportunities and challenges of nanotechnology-enhanced orthopedic materials and provide a comprehensive summary on the state of the art in each category of the materials.

References

[1] Feynman R. There's plenty of room at the bottom. Caltech Eng Sci 1960;23:22–36.
[2] Fanfair D, Desai S, Kelty C. The early history of nanotechnology, http://cnx.org/content/m14504/latest/; 2005 [accessed 21.01.13].
[3] Luttge R. Nanotechnology. In: Luttge R, editor. Microfabrication for industrial applications. Boston: William Andrew; 2011. p. 91–146.
[4] Barth JV, Costantini G, Kern K. Engineering atomic and molecular nanostructures at surfaces. Nature 2005;437:671–9.
[5] Lin BJ. Optical lithography: here is why. Bellinghan, WA: SPIE Press; 2010.
[6] Pease RF, Chou SY. Lithography and other patterning techniques for future electronics. Proc IEEE 2008;96:248–70.
[7] Gates BD, Xu Q, Stewart M, Ryan D, Willson CG, Whitesides GM. New approaches to nanofabrication: molding, printing, and other techniques. Chem Rev 2005;105:1171–96.
[8] Loo Y-L, Willett RL, Baldwin KW, Rogers JA. Additive, nanoscale patterning of metal films with a stamp and a surface chemistry mediated transfer process: applications in plastic electronics. Appl Phys Lett 2002;81:562–4.

[9] Piner RD, Zhu J, Xu F, Hong S, Mirkin CA. "Dip-Pen" nanolithography. Science 1999;283:661–3.
[10] Liu M, Ji Z, Shang L. Top-down fabrication of nanostructures. In: Chi L, editor. Nanotechnology. Weinheim: Willey-VCH Verlag GmbH & Co.; 2010. p. 3–47.
[11] Knez M, Nielsch K, Niinistö L. Synthesis and surface engineering of complex nanostructures by atomic layer deposition. Adv Mater 2007;19:3425–38.
[12] Chow L, Wang H, Kleckley S, Daly TK, Buseck PR. Fullerene formation during production of chemical vapor deposited diamond. Appl Phys Lett 1995;66:430–2.
[13] Fukui T, Ando S, Tokura Y, Toriyama T. GaAs tetrahedral quantum dot structures fabricated using selective area metalorganic chemical vapor deposition. Appl Phys Lett 1991;58:2018–20.
[14] Brunhes T, Boucaud P, Sauvage S, Aniel F, Lourtioz J-M, Hernandez C, et al. Electroluminescence of Ge/Si self-assembled quantum dots grown by chemical vapor deposition. Appl Phys Lett 2000;77:1822–4.
[15] Kitamura M, Nishioka M, Oshinowo J, Arakawa Y. In situ fabrication of self-aligned InGaAs quantum dots on GaAs multiatomic steps by metalorganic chemical vapor deposition. Appl Phys Lett 1995;66:3663–5.
[16] Kumar M, Ando Y. Chemical vapor deposition of carbon nanotubes: a review on growth mechanism and mass production. J Nanosci Nanotechnol 2010;10:3739–58.
[17] Koziol K, Boskovic BO, Yahya N. Synthesis of carbon nanostructures by CVD method. In: Yahya N, editor. Carbon and oxide nanostructures. Berlin, Heidelberg: Springer-Verlag; 2010.
[18] Cassell AM, Raymakers JA, Kong J, Dai H. Large scale CVD synthesis of single-walled carbon nanotubes. J Phys Chem B 1999;103:6484–92.
[19] Wu J-J, Liu S-C. Catalyst-free growth and characterization of ZnO nanorods. J Phys Chem B 2002;106:9546–51.
[20] Cui Y, Lauhon LJ, Gudiksen MS, Wang J, Lieber CM. Diameter-controlled synthesis of single-crystal silicon nanowires. Appl Phys Lett 2001;78:2214.
[21] Gruen DM. Nanocrystalline diamond films. Annu Rev Mater Sci 1999;29:211–59.
[22] Reina A, Jia X, Ho J, Nezich D, Son H, Bulovic V, et al. Large area, few-layer graphene films on arbitrary substrates by chemical vapor deposition. Nano Lett 2008;9:30–5.
[23] Zarrabian M, FourchesCoulon N, Turban G, Marhic C, Lancin M. Observation of nanocrystalline diamond in diamondlike carbon films deposited at room temperature in electron cyclotron resonance plasma. Appl Phys Lett 1997;70:2535–7.
[24] Pursel S, Horn MW, Demirel MC, Lakhtakia A. Growth of sculptured polymer submicronwire assemblies by vapor deposition. Polymer 2005;46:9544–8.
[25] Fujita J, Ishida M, Sakamoto T, Ochiai Y, Kaito T, Matsui S. Observation and characteristics of mechanical vibration in three-dimensional nanostructures and pillars grown by focused ion beam chemical vapor deposition. J Vac Sci Technol B: Microelectron Nanometer Struct 2001;19:2834–7.
[26] Susi T, Nasibulin AG, Jiang H, Kauppinen EI. CVD synthesis of hierarchical 3D MWCNT/carbon-fiber nanostructures. J Nanomater 2008;2008:61.
[27] Xu W, Ye Z, Zhu L, Zeng Y, Jiang L, Zhao B. ZnO nanostructure networks grown on silicon substrates. J Cryst Growth 2005;277:490–5.
[28] Lukaszkowicz K. Review of nanocomposite thin films and coatings deposited by PVD and CVD technology. In: Rahman M, editor. Nanomaterials. InTech; 2011. http://dx.doi.org/10.5772/25799, ISBN: 978-953-307-913-4. Available from: http://www.intechopen.com/books/nanomaterials/review-of-nanocomposite-thin-films-and-coatings-deposited-by-pvd-and-cvd-technology.

[29] Knez M, Kadri A, Wege C, Gösele U, Jeske H, Nielsch K. Atomic layer deposition on biological macromolecules: metal oxide coating of tobacco mosaic virus and ferritin. Nano Lett 2006;6:1172–7.
[30] Johnson D. Nanofabrication top down to bottom up. Online media, www.matecnetworks.org; 2011 [accessed 02.02.13].
[31] Yokoyama T, Huang C. Nanoparticle technology for the production of functional materials. Kona 2005;23:7–17.
[32] Atomization. Nanodiccom. http://www.nanodic.com/nanofabrication/Atomization.htm [accessed 02.02.13].
[33] Reverchon E. Process for the production of micro and/or nano particles. Google Patents; 2007.
[34] Shimomura M, Sawadaishi T. Bottom-up strategy of materials fabrication: a new trend in nanotechnology of soft materials. Curr Opin Colloid Interface Sci 2001;6:11–6.
[35] Yang L, Webster TJ. Nanotechnology controlled drug delivery for treating bone diseases. Expert Opin Drug Del 2009;6:851–64.
[36] Braun E, Eichen Y, Sivan U, Ben-Yoseph G. DNA-templated assembly and electrode attachment of a conducting silver wire. Nature 1998;391:775–8.
[37] Yan H, Park SH, Finkelstein G, Reif JH, LaBean TH. DNA-templated self-assembly of protein arrays and highly conductive nanowires. Science 2003;301:1882–4.
[38] Sharma J, Chhabra R, Liu Y, Ke Y, Yan H. DNA-templated self-assembly of two-dimensional and periodical gold nanoparticle arrays. Angew Chem Int Ed 2006;45:730–5.
[39] Cheng JY, Ross C, Thomas EL, Smith HI, Vancso GJ. Templated self-assembly of block copolymers: effect of substrate topography. Adv Mater 2003;15:1599–602.
[40] Cheng JY, Zhang F, Smith HI, Vancso GJ, Ross CA. Pattern registration between spherical block copolymer domains and topographical templates. Adv Mater 2006;18:597–601.
[41] Wang X, Zhuang J, Peng Q, Li Y. A general strategy for nanocrystal synthesis. Nature 2005;437:121–4.
[42] Caruso F, Caruso RA, Möhwald H. Nanoengineering of inorganic and hybrid hollow spheres by colloidal templating. Science 1998;282:1111–4.
[43] Xia Y, Yin Y, Lu Y, McLellan J. Template-assisted self-assembly of spherical colloids into complex and controllable structures. Adv Funct Mater 2003;13:907–18.
[44] Cheng JY, Ross CA, Smith HI, Thomas EL. Templated self-assembly of block copolymers: top-down helps bottom-up. Adv Mater 2006;18:2505–21.
[45] Gashti MP, Willoughby J, Agrawal P. Surface and bulk modification of synthetic textiles to improve dyeability. In: Hauser P, editor. Textile dyeing. InTech; 2011. http://dx.doi.org/10.5772/18706, ISBN: 978-953-307-565-5. Available from: http://www.intechopen.com/books/textile-dyeing/surface-and-bulk-modification-of-synthetic-textiles-to-improve-dyeability.
[46] Kim IS, Kumta PN. Sol–gel synthesis and characterization of nanostructured hydroxyapatite powder. Mater Sci Eng B-Solid 2004;111:232–6.
[47] Mathers JP, Wood WP. Aluminum nitride/aluminum oxynitride/group IVB metal nitride abrasive particles derived from a sol–gel process. Google Patents; 1988.
[48] Bigi A, Boanini E, Bracci B, Facchini A, Panzavolta S, Segatti F, et al. Nanocrystalline hydroxyapatite coatings on titanium: a new fast biomimetic method. Biomaterials 2005;26:4085–9.
[49] Borcia G, Chiper A, Rusu I. Using a He + N2 dielectric barrier discharge for the modification of polymer surface properties. Plasma Sources Sci Technol 2006;15:849.

[50] Yang L, Li YW, Sheldon BW, Webster TJ. Altering surface energy of nanocrystalline diamond to control osteoblast responses. J Mater Chem 2012;22:205–14.
[51] Utegulov ZN, Mast DB, He P, Shi D, Gilland RF. Functionalization of single-walled carbon nanotubes using isotropic plasma treatment: resonant Raman spectroscopy study. J Appl Phys 2005;97:104324-1–4.
[52] Weng C, Leou K, Wei H, Juang Z, Wei M, Tung C, et al. Structural transformation and field emission enhancement of carbon nanofibers by energetic argon plasma post-treatment. Appl Phys Lett 2004;85:4732–4.
[53] Gokus T, Nair R, Bonetti A, Bohmler M, Lombardo A, Novoselov K, et al. Making graphene luminescent by oxygen plasma treatment. ACS Nano 2009;3:3963–8.
[54] Chu PK, Chen J, Wang L, Huang N. Plasma-surface modification of biomaterials. Mat Sci Eng: R 2002;36:143–206.
[55] Hu J, Shao D, Chen C, Sheng G, Li J, Wang X, et al. Plasma-induced grafting of cyclodextrin onto multiwall carbon nanotube/iron oxides for adsorbent application. J Phys Chem B 2010;114:6779–85.
[56] Oksuz AU, Manolache S, Oksuz L, Hershowitz N. Plasma nanocoating of thiophene onto TiO_2 nano particles. Ind Eng Chem Res 2013; 52(19): 6610–6.
[57] Bidan G. Giving nano-objects new properties by molecular grafting. In: Center CG, editor. CLEFS CEA. 2005. p. 67–74. Online media, http://www.cea.fr/english-portal/library/clefs-cea/clefs-cea-no.-52-the-nanoworld-from-the-labora [accessed 02.02.13].
[58] Liu Q, de Wijn JR, de Groot K, van Blitterswijk CA. Surface modification of nano-apatite by grafting organic polymer. Biomaterials 1998;19:1067–72.
[59] Veronese FM, Pasut G. PEGylation, successful approach to drug delivery. Drug Discov Today 2005;10:1451–8.
[60] Machado MC, Tarquinio KM, Webster TJ. Decreased *Staphylococcus aureus* biofilm formation on nanomodified endotracheal tubes: a dynamic airway model. Int J Nanomed 2012;7:3741.
[61] Rho JY, Kuhn-Spearing L, Zioupos P. Mechanical properties and the hierarchical structure of bone. Med Eng Phys 1998;20:92–102.
[62] Puleo D, Nanci A. Understanding and controlling the bone–implant interface. Biomaterials 1999;20:2311–21.
[63] Anderson JM, Gristina AG, Hanson SR, Harker LA, Johnson RJ, Merritt K, et al. Host reactions to biomaterials and their evaluation. In: Ratner BD, Hoffman AS, Schoen FJ, Lemons JE, editors. Biomaterials science: an introduction to materials in medicine. San Diego: Academic Press; 1996. p. 165–213.
[64] Wilson CJ, Clegg RE, Leavesley DI, Pearcy MJ. Mediation of biomaterial–cell interactions by adsorbed proteins: a review. Tissue Eng 2005;11:1–18.
[65] Liu H, Webster TJ. Nanomedicine for implants: a review of studies and necessary experimental tools. Biomaterials 2006;28:354–69.
[66] Thomas KA, Cook SD. An evaluation of variables influencing implant fixation by direct bone apposition. J Biomed Mater Res 1985;19:875–901.
[67] Price RL, Waid MC, Haberstroh KM, Webster TJ. Selective bone cell adhesion on formulations containing carbon nanofibers. Biomaterials 2003;24:1877–87.
[68] Washburn NR, Yamada KM, Simon Jr. CG, Kennedy SB, Amis EJ. High-throughput investigation of osteoblast response to polymer crystallinity: influence of nanometer-scale roughness on proliferation. Biomaterials 2004;25:1215–24.
[69] Webster TJ, Hellenmeyer EL, Price RL. Increased osteoblast functions on theta plus delta nanofiber alumina. Biomaterials 2005;26:953–60.

[70] Keselowsky BG, Collard DM, Garcia AJ. Integrin binding specificity regulates biomaterial surface chemistry effects on cell differentiation. Proc Natl Acad Sci U S A 2005;102:5953–7.

[71] Balasundaram G. Nanomaterials for better orthopedics. In: Webster TJ, editor. Nanotechnology for the regeneration of hard and soft tissues. Hackensack, NJ; London: World Scientific; 2007. p. 53–78.

[72] Zhu XS, Zhu L, Duan ZH, Qi RQ, Li Y, Lang YP. Comparative toxicity of several metal oxide nanoparticle aqueous suspensions to Zebrafish (*Danio rerio*) early developmental stage. J Environ Sci Heal A 2008;43:278–84.

[73] McHale G, Shirtcliffe NJ, Aqil S, Perry CC, Newton MI. Topography driven spreading. Phys Rev Lett 2004;93:036102.

[74] Discher DE, Janmey P, Wang YL. Tissue cells feel and respond to the stiffness of their substrate. Science 2005;310:1139–43.

[75] Chan CE, Odde DJ. Traction dynamics of filopodia on compliant substrates. Science 2008;322:1687–91.

[76] Pelham Jr. RJ, Wang Y. Cell locomotion and focal adhesions are regulated by substrate flexibility. Proc Natl Acad Sci U S A 1997;94:13661–5.

[77] Qiu Q, Sayer M, Kawaja M, Shen X, Davies JE. Attachment, morphology, and protein expression of rat marrow stromal cells cultured on charged substrate surfaces. J Biomed Mater Res 1998;42:117–27.

[78] Itoh S, Nakamura S, Nakamura M, Shinomiya K, Yamashita K. Enhanced bone ingrowth into hydroxyapatite with interconnected pores by electrical polarization. Biomaterials 2006;27:5572–9.

Nanotechnology-enhanced metals and alloys for orthopedic implants

2

Metals and alloys have been extensively used for manufacturing prosthetic implants and demonstrated to be clinically safe and effective in orthopedic applications since the beginning of the twentieth century. For years, metallic or alloy implants (including those with ceramic or polymeric components) have been the most common (or sometimes the only) choices for the surgical treatment of patients suffered from hip, knee, spinal, elbow fractures or other problems. This prevalence of use places metallic materials in an important position in the development of new and better orthopedic implants. Over the past two decades, the emergence of nanotechnology has revolutionized the research and development of metallic medical implants, fostering their development from primarily structural materials that provide mechanical support and stability toward materials with both enhanced mechanical and biological properties. These enhanced properties include, but are not limited to, strength, superplasticity, wear resistance, corrosion resistance, biocompatibility, and osteogenic or osseointegration properties. Therefore, nanotechnology-enhanced metals and alloys have received lots of attention from the orthopedic community, and a number of commercialized applications are expected to enter the market in the near future. In this chapter, fabrication techniques, approaches to improve mechanical properties, studies on controlling cell and protein responses, and commercialization status of nanostructured metals and alloys for orthopedic applications are introduced. In particular, Ti-based alloys and anodized Ti for orthopedic applications are emphasized. In addition, novel nanostructured metals and alloys possessing attractive functionalities such as anti-infection, drug delivery, biosensing and detection, and superior mechanical properties are also covered.

2.1 Fabrication techniques of nanostructured metals and alloys

Manufacture of nanostructured or nanograined metals and alloys mainly relies on two categories of approaches, one for producing fine powders (i.e., particulates) and the other for bulk materials. To date, numerous fabrication and synthesis methods of nanostructured metals in both particulate and bulk forms have been developed, but many of them are still distant from industrial-scale production. This section focuses on the fabrication and processing techniques that have high potential for industrial manufacture.

Nanotechnology-Enhanced Orthopedic Materials. http://dx.doi.org/10.1016/B978-0-85709-844-3.00002-1

2.1.1 Vapor condensation or deposition

Vapor condensation or deposition is one of the most maturely developed techniques for synthesizing nanostructured metals in which metal atoms condense from vapor phase [1]. Vapor condensation and vapor deposition (i.e., the physical vapor deposition and/or chemical vapor deposition introduced in Chapter 1) techniques are similar except the final products from vapor condensation are particulate while the vapor deposition usually yields thin films. Both techniques have been used for synthesizing a broad range of metals with nanosized or submicron-sized grains (or crystals), including Fe, Co, Al, Ti, Cr, Ni, Mg, Ta, Au, and Ag as well as alloys based on these metals [2–5]. Vapor condensation and deposition techniques have also been developed to produce complex architectures or superstructures of metals and alloys on nanoscale, such as core–shell structure, capsule, and multilayers. For instance, nanocapsules of Fe and Co coated by carbon are produced by chemical vapor condensation (CVC) using organometallic precursors (e.g., iron carbonyl and cobalt carbonyl) and a carrier gas of carbon monoxide [6]. In addition, specific properties of nanometals or alloy can be controlled during the CVC process. For example, Fe nanoparticles with different magnetic states from superparamagnetic to ferromagnetic can be fabricated by altering CVC processing parameters [7].

2.1.2 High-energy ball milling

High-energy ball milling is a mechanical deformation process that is frequently used for producing nanocrystalline metals or alloys in powder form. This technique belongs to the comminution or attrition approach introduced in Chapter 1. In the high-energy ball milling process, coarse-grained structures undergo disassociation as the result of severe cyclic deformation induced by milling with stiff balls in a high-energy shaker mill [8,9]. This process has been successfully used to produce metals with minimum particle sizes from 4 to 26 nm. The high-energy ball milling technique is simple and has high potential to scale up to produce tonnage quantities of materials [8]. However, a serious problem of this technique is the contamination from milling media (balls and vial) and/or atmosphere. Therefore, a number of improvements, including the usages of surfactants, alloy-coated milling media, and protective atmospheres, have been developed to alleviate the contamination problem [8].

The fine powder (in nano or submicron sizes) produced from ball milling can be consolidated to bulk form for large-scale applications such as hip implants and bone screws. Usually, the fine powders are compacted and sintered together via methods like hot isostatic pressing and explosive compaction under the temperatures or conditions that suppress grain growth and maintain nanocrystalline microstructure [8,10]. Bulk metallic materials produced by this approach have achieved the theoretical densities of nanocrystalline materials and greatly improved mechanical properties compared to their conventional, micron-grained counterparts.

2.1.3 Wet-chemical synthesis

As introduced in Chapter 1, a number of wet-chemical, bottom-up synthesis techniques are effective in producing metallic or alloy nanoparticles. The representatives of

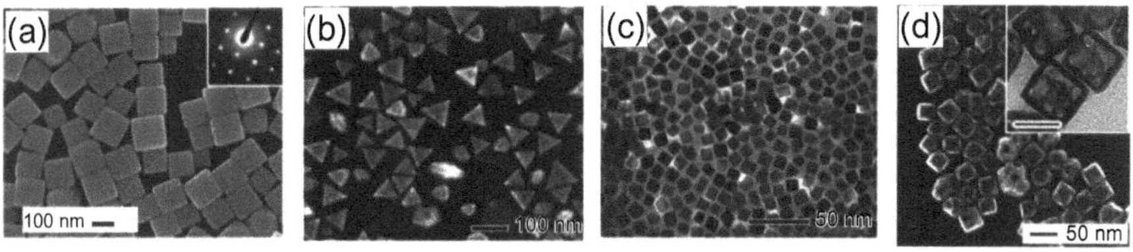

Figure 2.1 Examples of metallic nanoparticles synthesized by wet-chemical methods: (a) Ag nanocubes, (b) nano-Au truncated tetrahedrons, (c) Pt nanocubes, and (d) Au/Ag/Pd alloy nanocages.
Created with permission from [11,12].

nanoparticles synthesized by these techniques are noble metals including Au, Ag, Pt, and their alloys, some of which reveal considerable potential in orthopedic applications. Compared to other synthesis techniques, wet-chemical synthesis is more precise in controlling nanoparticle size and shape. As illustrated in Figure 2.1, several types of metallic nanoparticles with high uniformities and narrow size distributions (also known as mono-dispersed nanoparticles) can be synthesized by galvanic replacement reaction between solutions containing metal precursor salts and Ag nanostructures prepared through polyol reduction [11,12]. In addition, the subtle shape control of metallic nanoparticles can also be achieved during the wet-chemical synthesizing process via precise controls of reaction kinetics and/or thermodynamics [12]. Figure 2.2 illustrates an example of gradually modifying Au nanocubes to obtain porous nanocubes and eventually nanoframes by wet etching [11], revealing the versatility of wet-chemical synthesis in fabricating nanometals with various geometries.

2.1.4 Severe plastic deformation

For fabricating bulk metals and alloys with nanometer grains or phases, severe plastic deformation (SPD) techniques are efficient choices with high reproducibility. Typically, in order to refine the microstructures of bulk metals, SPD techniques employ large-scale deformation processes, such as rolling, drawing, equal channel angular pressing/extrusion (ECAP/E), or high-pressure torsional strain [10]. In the SPD processes, extreme plastic strains (typically >4 in total) are imposed in a bulk metal through repeated application of plastic deformation, and the initial grain sizes (usually in micron range) are reduced to nanometer or submicron regimes after multiple times of deformation [10,13]. Figure 2.3(a) illustrates the principles of ECAP, in which a metal billet is fed into a die and pressed by a plunger to pass an equal but angled channel for multiple times. In each pass, the cross-sectional area of the metal remains unchanged and the substantial strain imposed in the metal is a function of the channel angles [14]. Figure 2.3(b) shows the electron microscopy images of EACP-processed Mg alloy after eight passes at 250 °C and four passes at 150 °C, where its ultrafine grains with sizes of a few hundred nanometers can be observed.

SPD techniques like ECAP share several advantages including high potential for scaling up, high reproducibility, and notable effectiveness in producing nanostructured bulk materials from ductile metals and alloys of initially low to moderate strengths.

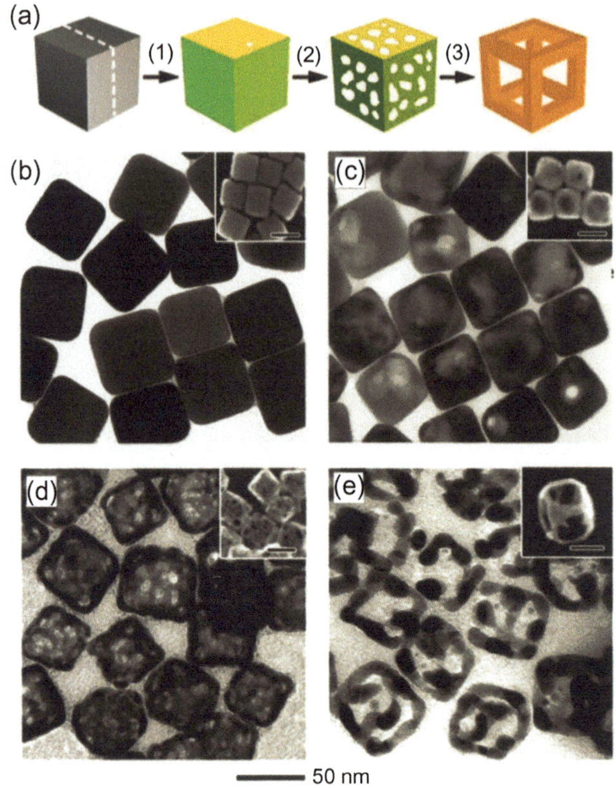

Figure 2.2 (a) Illustration showing the three steps of cubic Au nanoframe formation: (1) beginning with Ag nanocube (TEM shown in (b)), Ag/Au nanobox formed by galvanic replacement (TEM shown in (c)); (2) wet etchant removes remaining Ag to form a porous nanocage (TEM shown in (d)); (3) more etching transferring the nanocage to a nanoframe (TEM shown in (e)).
Reprinted with permission from Ref. [11].

However, SPD techniques appear to be difficult in processing metals and alloys possessing initially high strengths (i.e., difficult to deform) and the equipment for large volume production may be expensive [10].

2.1.5 Anodization

Nanoporous surfaces can be readily fabricated on some metals by potentiostatic or galvanostatic anodization of the metals in acidic electrolytes (e.g., H_2SO_4, HNO_3, HF, etc.). These metals include Al and several valve metals such as Ti, Ta, and Nb [15]. In a typical anodization cell (as illustrated in Figure 2.4a), Al or valve metal acts as anode and platinum is usually used as cathode and both are immersed in an acidic electrolyte. When a proper electric field is applied to the electrochemical cell, nanostructured oxide can form on the anodic metal. Specifically, the anodization technique can create uniform nanotubular structures on Ti (as shown in Figure 2.4b) and Ta surfaces with diameters

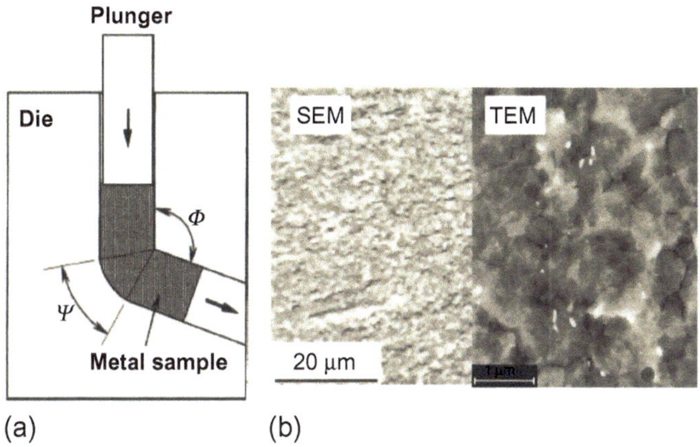

Figure 2.3 Illustrations of equal channel angular pressing (ECAP) technique: (a) schematic of ECAP process and (b) scanning electron microscopy (SEM) and transmission electron microscopy (TEM) images showing nanostructures of Mg after ECAP processing.
Scale bar = 1 micron in the TEM image.
(a) was modified with permission from Ref. [14], and (b) from personal communication with Dr. Maurizio Vedani.

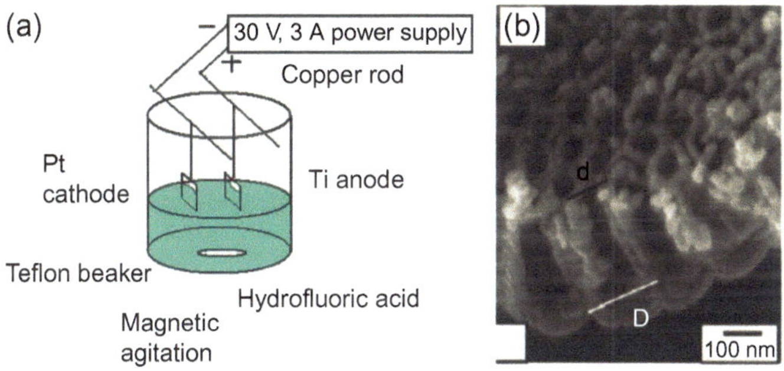

Figure 2.4 Illustration of Ti anodization to create nanoscale tubular oxide layer on metal surface: (a) experimental setup of anodization and (b) SEM image of Ti nanotube arrays after anodization. Images modified with permission from Dr. Thomas J. Webster.

less than 100 nm [16,17]. Due to the simplicity and reproducibility of the process, anodization technique has been widely used for producing nanoporous Ti-based medical implants [18]. Strictly speaking, the resultant nanostructured materials from anodization are metal oxides (i.e., ceramics) rather than metals or alloys. Nevertheless, a number of anodized oxides are discussed in this chapter because of their close connections to orthopedic metals and alloys.

2.1.6 Other fabrication methods

A regular metal removal process, lathe machining, has been shown to be a possible method for producing a wide group of nanocrystalline metals and alloys at very low cost. A study by Brown et al. has demonstrated that the chips formed from machining metals and alloys were composed of nanocrystalline domains of high hardness, probably due to large strain deformation during the machining [10]. This study provides a simple avenue for manufacturing nanostructured metals and alloys (in forms of chips or particulates) in large volume from the chips or scraps generated in industrial machining operations. Compared to the techniques mentioned before, this technique has significant economic and environmental benefits.

Plasma spaying is a representative of spray processing approaches for fabricating nanometer materials or coatings. Plasma spraying is an advanced but mature technique that has recently been used for depositing nanocrystalline metals and alloys as well as ceramics and oxides [19]. In the plasma spray process, a high temperature plasma jet is used to melt and/or spray a feedstock (e.g., liquid precursor), which is injected into the plasma flame. The feedstock flows and then precipitates on a collector or deposits on a substrate to form nanosized particles or coatings [19,20]. By far, nanocrystalline metals and alloys including Al alloys, WC–Co coatings, TiO_2/Ti alloys, and TiO_2/Ag alloys have been produced by plasma spraying for potential orthopedic applications [21–24].

Several surface-roughening strategies have also been applied to modify bulk metals and alloys in order to create nanometer topological surface features. Grit blasting is one of the well-established surface-roughening techniques for metals such as Ti and its alloys, steel, and CoCrMo [15,25]. Grit blasting is a process in which angular-shaped metallic or ceramic grits are carried by a pressurized gas stream and hurled against the surface of a working material. The sharp grits of proper sizes erode the surface to generate nanometer roughness or form nanocrystalline regions (usually in a depth of a few microns) by work-hardening effects [25]. Ultrafast laser irradiation is a recently developed technique for modifying metal surfaces to obtain nanostructures [26]. In this process, an ultrafast laser pulse interacts with metallic materials to form self-organized conical nanostructures on various metal surfaces like Ti, Al, Cu, and stainless steel. Furthermore, the size and geometry of such nanostructures can be controlled by laser parameters. For example, highly ordered arrays of micro-/nanoholes with uniform sizes could be formed in thin Ti foils by this laser treatment [26].

2.2 Nanostructured metals for better orthopedic implants with improved biological functions

2.2.1 *Ti-based biomaterials*

Ti-based biomaterials (such as commercial pure titanium (cpTi), Ti–6Al–4V, and Ti–6Al–7Nb) have been widely used for orthopedic, dental, and maxillofacial applications over the past few decades [27]. Ti and its alloys have high strength-to-weight ratios, proper stiffness, high load resistance and fatigue resistance, sufficient ductility for physiological loading, and excellent biocompatibility properties. Titanium oxides, formed on Ti surfaces in oxygen-containing environments, are also biocompatible and provides a kinetic barrier that prevents corrosion. However, the functional clinical lifetime of current Ti prostheses/implants is approximately 15–20 years [28]. In many failure cases, bone integration with implant is either not achieved or once established, it can be lost due to scar tissue formation, wear debris generation, infection, and so on [29,30]. Therefore, promoting bone cell functions to enhance bone integration (i.e., osseointegration) at the Ti implant surface becomes extremely important.

On one hand, bone cell functions are sensitive to the chemical and structural features on Ti surfaces. For example, syntheses of autocrine/paracrine regulatory chemicals (such as prostaglandin E_2), transforming growth factor beta-1, and osteoprotegerin increases when osteoblasts (bone-forming cells) are cultured on Ti with increased nano- to micron-surface features [31]. On the other hand, on nanosmooth, microrough surfaces more fibrous callus formation is observed and this outcome is undesirable for Ti implant efficacy. With this in mind, Ti and its alloys can be engineered in surface chemistry, charge, wettability, and topography to control protein adsorption and subsequent cell adhesion and other functions [32]. Ti surface treatments, such as chemical etching and anodization, have been developed and investigated in the orthopedic field to increase osseointegration and biocompatibility of the implants [33,34].

In particular, anodization technique can create uniform pores or tubes on the surfaces of Ti and Ti alloys with diameters less than 100 nm and lengths of a few hundred nanometers, as shown in Figure 2.5. The nanometer surface roughness from such structures has shown to improve protein adsorption (such as vitronectin and fibronectin), subsequent osteoblast attachment, and eventual osseointegration. For example, Huang et al. anodized Ti in a 5 M sodium hydroxide (NaOH) solution at 25 °C with a potentiodynamic anodization treatment and reported that the number of osteoblast-like cells adhering on TiO_2/Ti_2O_3 increased more than 1.4 times when compared to untreated Ti [36]. Ti anodization in 1.5 wt.% hydrofluoric acid (HF) at 20 V for 10 s to 5 min also promoted osteoblast adhesion [33]. In other studies, osteoblast density, proliferation, and differentiation (osteoblasts transform from proliferation to bone-forming status) also revealed significant increases on anodized Ti surfaces compared to untreated Ti [37,38]. These enhancements have been attributed to nanoscale surface roughness and higher surface energy or wettability of anodized Ti compared to conventional Ti.

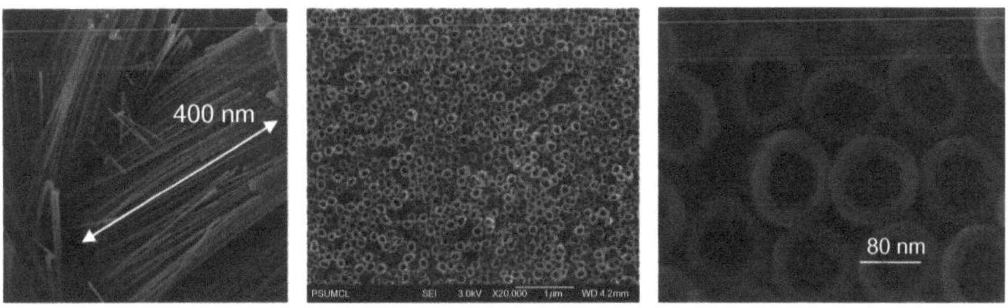

Figure 2.5 SEM images of anodized Ti nanotubular surfaces. Left: Cross-sectional view of mechanically fractured sample showing that the length of the tubes was approximately 400 nm; Center: top-view of nanotubular surface; Right: high-magnification top-view of a nanotubular surface showing tube diameters of approximately 80 nm.
Reproduced from Ref. [35] with permission.

The morphology of osteoblast precursor cells adhering on control Ti and anodized Ti, which were created nanoporous in the electrolyte solutions containing either H_2SO_4, H_3PO_4, or HF, after 11 days are shown in Figure 2.6. More cell attachment and increased cell–cell interactions were observed on the anodized Ti, also revealing more focal adhesion of cell onto anodized Ti. Alkaline phosphatase activity (an important marker indicating osteoblast differentiation status) of osteoblasts after 7 days and calcium deposition in their extracellular matrix after 3 weeks were both 50% higher on nanotubular surfaces (anodized in 0.5% HF for 45 min with 20 V creating nanotubes approximately 80 nm in diameter) compared to nonanodized surfaces *in vitro* [35]. Animal studies on anodized Ti also testify to the efficacy of Ti nanostructures in promoting bone growth. Figure 2.7 reveals the histological results of bone growths on untreated and anodized Ti screws in amputee rat models after 28 days of implantation, respectively. The results clearly revealed that new bone growth with anodized Ti screws was much more substantial than that with untreated Ti screws.

Nanorough Ti fabricated by other fabrication processes also lead to improved osteoblast functions. For instance, de Oliverira et al. reported that nanotextured Ti, which was prepared by chemical etching, enhanced metabolic activities of cultured bone cells (i.e., up-regulations of bone sialoprotein and osteopontin) [39]. Similarly,

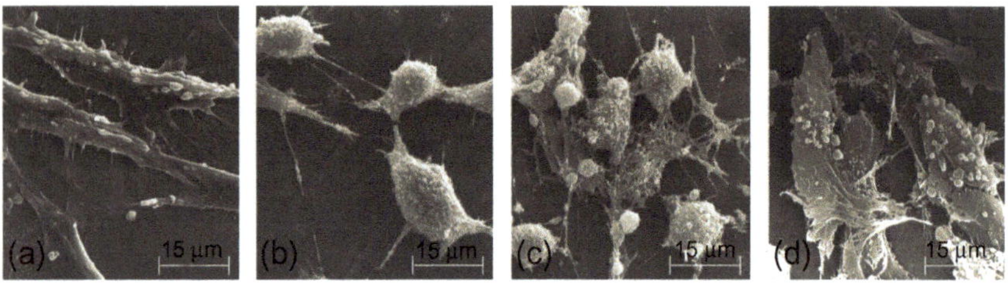

Figure 2.6 SEM micrographs illustrating the adhesion and proliferation of osteoblastic precursor cell lines (OPC1), established from human fetal bone tissue, on (a) control Ti, anodized Ti (thus, creating nanometer surface features) in (b) H_2SO_4, (c) H_3PO_4, and (d) HF electrolytes after 11 days of culture.
Modified with permission from Ref. [38].

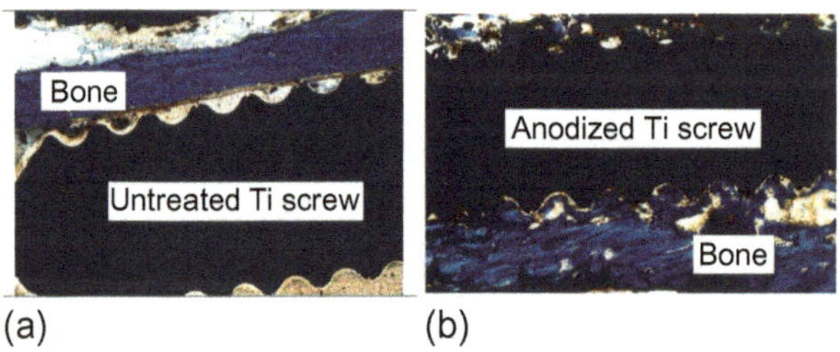

Figure 2.7 Histological results from rat amputee models showing (a) lack of bone (blue stain) growth with untreated Ti screw and (b) increased bone growth with anodized Ti screw, after 28 days of implantation.
Modified with permission from Prof. Thomas J. Webster.

an ionic plasma deposition process that deposited nanostructured Ti on silicon substrates increased osteoblast but decreased fibroblast adhesion when compared with bare silicon and Ti *in vitro* [40]. Nanoparticles of cpTi and prealloyed Ti–6Al–4V, pressurized at 10 GPa over 5 min into compacts for testing, also revealed the promoted osteoblast adhesion, and this promotion was correlated to the increase in surface grain boundaries of metals [34]. In addition, Ward et al. found greater calcium and phosphorus mineral contents in the extracellular matrix of osteoblasts cultured on nanophase Ti–6Al–4V that was also compacted from nanophase powders [41]. They reported that nanophase surfaces of Ti-based implants promoted osteoblast adhesion up to three times compared to nonanodized Ti, revealing the positive role of nanometer roughness to cell–biomaterial interactions.

2.2.2 Nanophase CoCrMo

Similar to nanophase Ti and Ti alloys, nanophase CoCrMo has also shown to promote bone growth *in vitro*. For example, nanophase CoCrMo compacts prepared by pressurizing Co, Cr, Mo elemental nanoparticles at 5 GPa over 5 min have demonstrated significant promotion of osteoblast adhesion [34]. The similar nanophase CoCrMo compacts have also revealed increased calcium and phosphorus mineral depositions in the extracellular matrix of attached osteoblasts [41]. The strategy of anodization has also been applied to CoCrMo to generate nanorough surfaces for promoting bone growths. Chang and Webster et al. demonstrated that anodization of clinical CoCrMo in HF at 10 V yielded rough oxide surfaces composed of both nano- and microporous structures and such rough surfaces resulted in 24% more osteoblast adhesion after 4 h compared to conventional CoCrMo surfaces [42]. The promotion of bone cell functions on nanostructured CoCrMo was attributed to greatly increased surface area, grain boundary phase and, consequently, an elevated amount of adhesion proteins (e.g., fibronectin, vitronectin, etc.) on nanophase metals compared to convention, microstructured metals.

Besides promoting protein adsorption and cell functions, nanostructured CoCrMo may help reduce wear debris from the counterface of ultrahigh molecular weight polyethylene (UHMWPE) and, therefore, help suppress adverse host responses from surrounding tissues. For instance, nanotextured CoCrMo surface was developed to minimize the amount of wear of UHMWPE and to increase the size of UHMWPE wear particles, which can reduce macrophage activation that eventually causes bone resorption [43]. The nanotextured CoCrMo surface had a superfine surface with 3% of groove and dimples against the bearing area, and the depths of groove and dimples are less than 50 nm. Wear tests revealed that the nanotextured surface reduced the amount of UHMWPE wear compared to the conventional CoCrMo surface. More importantly, the wear debris generated on the nanotextured surface inhibited the genetic expression of IL-6 in cells (RAW264.7, blood, mouse) and could, thus, reduce macrophage activation, tissue reaction, and subsequent joint loosening.

However, nanostructured CoCrMo implants still possess a potential risk of long-term toxicity caused by nanoscale wear debris from the articulating surfaces [30], especially in the case of metal-on-metal implants. In a recent study, cytotoxic and genotoxic effects of nanoparticles and micron-sized particles of CoCr alloy were compared using human fibroblasts in tissue culture [44]. Results indicated that the nanoparticles, which caused more free radicals in an acellular environment, induced more DNA damage and subsequent cytotoxicity than the micron-sized particles at equivalent volumetric doses. Therefore, the concept of nanotoxicology for materials like nanophase CoCrMo becomes increasingly important in the design of future orthopedic implants. This topic will be further discussed in Chapter 9.

2.2.3 Nanostructured selenium (Se) for inhibiting cancer cell

Orthopedic implants are often used to replace resected cancerous bone but cancer recurrence near the implants remains a high probability. In this case, it would be beneficial to have implants specifically designed to prevent the recurrence of bone cancer and, simultaneously, promote healthy bone tissue growth. Recently, a novel idea of integrating conventional implant surfaces with anticancer chemistry via nanotechnology has been developed. Phong et al. created nanostructured surfaces for promoting healthy bone tissue growth and preventing carcinogenesis by incorporating Se nanoclusters on the implant material surfaces [45,46]. Se is an element naturally existing in human body and it is known to inhibit various types of cancer, including pancreatic, gastric, lung, nasopharyngeal, breast, uterine, respiratory, digestive, hematological, and gynecological cancers [47].

Se nanoclusters were synthesized and incorporated *in situ* on Ti, stainless steel, or CoCrMo substrates through a colloidal process [48]. The surfaces of Se-coated Ti and stainless steel are shown in Figure 2.8, and nanosized Se clusters

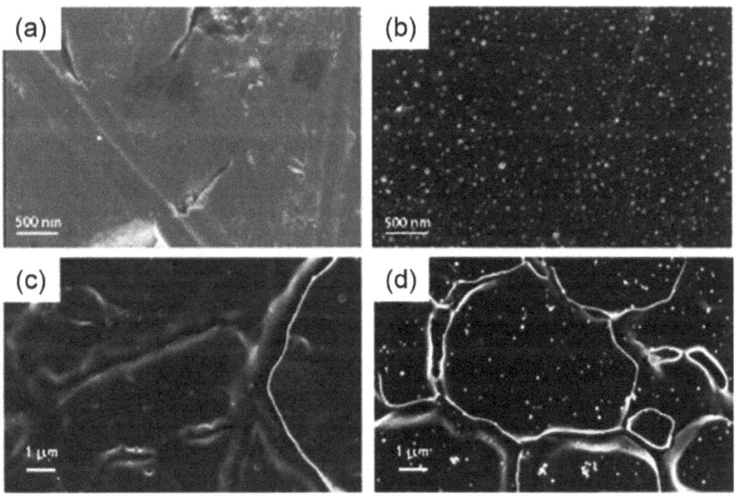

Figure 2.8 Scanning electron microscopy images of (a) uncoated Ti; (b) Se nanoclusters coated Ti; (c) uncoated stainless steel; and (d) Se nanoclusters coated stainless steel. Modified with permission from Ref. [45].

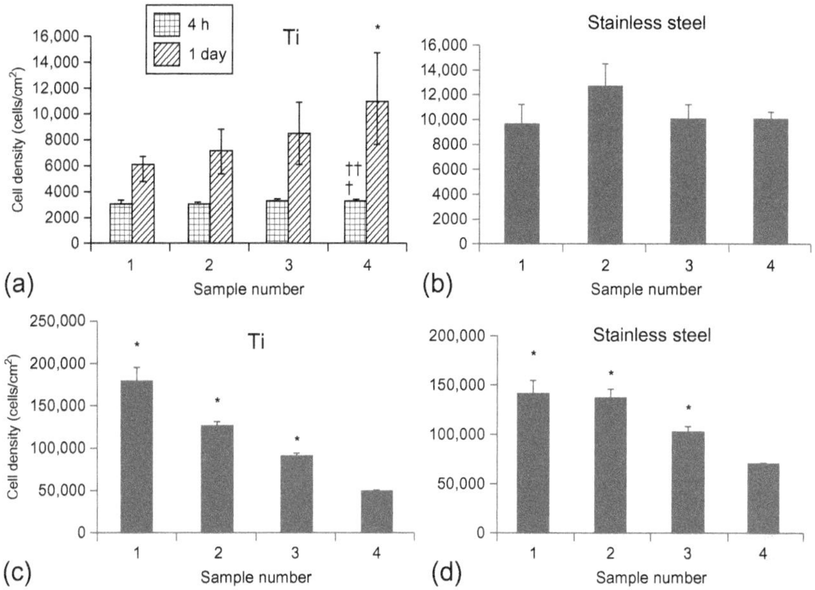

Figure 2.9 (a) Noncancerous osteoblast densities on various Se-coated Ti after 4 h and 1 day of culture; (b) noncancerous osteoblast densities on Se-coated stainless steel after 1 day; (c) cancerous osteoblast densities on Se-coated Ti, and (d) cancerous osteoblast densities on Se-coated stainless steel after 3-day culture. Sample number 1: uncoated metal; 2: low-content Se-coated metal; 3: medium-content Se-coated metal; 4: high-content Se-coated metal. Data = mean ± SEM; $N = 3$, *, † $p < 0.05$ compared to uncoated metals at same time period. †† $p < 0.05$ compared to low-content Se-coated metals at same time period.
Modified with permission from Ref. [45].

were strongly bonded to the metal surfaces. Implant surfaces coated with low (0.05 wt.%), medium (1.44 wt.%) and high (9.63 wt.%) doses of Se were investigated for the responses of noncancerous (i.e., healthy) and cancerous osteoblasts. Figure 2.9 illustrates the adhesion and proliferation of noncancerous and cancerous osteoblasts on Ti and stainless steel coated with Se. The results revealed an increase in noncancerous osteoblast density on Se-coated Ti compared to uncoated Ti but similar osteoblast densities on both coated and uncoated stainless steel. These results suggested that Se-coated implants are cytocompatible. However, the function of cancerous osteoblasts on these Se-coated surfaces was significantly inhibited compared to the results obtained on non-Se-coated implant materials (Figure 2.9). These results demonstrated that incorporation of nanophase Se on traditional implant materials (Ti and stainless steel) had no adverse effects on healthy osteoblast functions *in vitro* (adhesion and proliferation) but achieved an anticancer effect toward cancerous bone cells up to 3 days.

2.3 Nanotechnology-enabled functionality in metallic implants

2.3.1 Nanostructured metals for preventing infection

Bacterial infection is a common problem associated with orthopedic implantation procedures, causing serious complications in host tissues, failure of orthopedic implants, and even death of patients. Although antibiotics are widely used, problems of toxicity, antibiotic resistance, adverse responses of patients, short effective time, and availability always demand better approaches to prevent infection. Therefore, a new area of preventing infection has emerged by utilizing the extraordinary surface properties of nanomaterials (such as ultrasmall structures, increased surface area and roughness, increased grain boundaries, etc.) for antibiotic purposes.

A number of nanostructures have been studied for this purpose and revealed promising results to date. Puckett et al. examined *in vitro* adhesion of *Staphylococcus aureus (S. aures), Staphylococcus epidermidis (S. epidermidis)*, and *Pseudomonas aeruginosa (P. aeruginosa)* on conventional Ti, nanorough Ti (produced by electron beam evaporation), and nanotubular and nanotextured Ti (produced by two different anodization processes, respectively) [49]. The results indicated that the nanorough Ti surfaces decreased the adhesion of all of the aforementioned bacteria the most compared to conventional Ti (micron-sized grains), demonstrating the promise for fabricating anti-infection implants through nanotechnology. However, the nanotubular and nanotextured Ti revealed increased bacterial adhesions compared to conventional Ti. Similarly, increased *S. epidermidis* and *P. aeruginosa* attachments were reported on ultrafine-grained cpTi produced by ECAP compared to as-received cpTi [50]. Further studies indicate that antibacterial efficacy is largely influenced by chemical and structural parameters of nanostructured Ti, such as subtle chemistry change, diameter of nanotubes on anodized Ti, variable surface roughness and surface free energy, and crystallinity [49–51]. These results suggest that anti-infection efficacy of nanostructured Ti needs further study and understanding before clinical applications.

Another approach to improve the anti-infection efficacy of Ti or CoCrMo implants is incorporating antibacterial nanoparticles to the metallic implant surfaces. The typical antibacterial nanoparticles studied to date include Ag and Se. Ag has long been known as an inorganic agent killing bacteria, and the incorporation of Ag nanoparticles to anodized Ti nanotubes [52] and silanized CoCrMo [53] has been reported. For Ti nanotubes incorporated with Ag nanoparticles, the antibacterial rates were 100% in the first 10 days and the bacteria inhibitory effect of Ag nanoparticles lasted for as long as 30 days (antibacterial rates remained >80%). Antibacterial effects of Se against *S. aureus* and *S. epidermidis* biofilms have also been reported in several recent studies [54–56]. As mentioned in Section 2.2.3, Se nanoparticles can be readily coated on common orthopedic implant materials including Ti, CoCrMo, and stainless steel. A recent study by Holinka et al. reported that Ti discs coated with 0.5% and 0.2% Se (in forms of nano- and submicron particles) significantly reduced *S. aureus* and *S. epidermidis* attachment, while no inhibitory effect on osteoblastic cell growth was observed on these Se-coated Ti discs. Although more evaluations on their anti-infection and safety properties are needed, nanostructured Se and Ag are promising materials for reducing bacterial attachment on prosthetic implants [56].

2.3.2 Drug delivery via nanostructured implants

Although metals and alloys are popular orthopedic implant materials, they are usually not bioactive and do not exhibit any inherent bone-bonding capability. In order to improve new bone formation, reduce infection, and therefore increase implant lifetime *in vivo*, the strategy of combining drug delivery and implants via nanotechnology has been investigated. The drugs of interest include osteogenic growth factors, antibiotics and anti-inflammation agents, vaccines and nuclei acids [57]. Among these drugs, bone morphogenetic proteins (BMPs) and well-accepted antibiotics have received substantial attention, largely due to their therapeutic effects on osteoporosis and infection, respectively. BMPs originate from the transforming growth factor-β family (including at least 18 protein members) and can induce bone formation. It is widely recognized that BMP-2 possesses the highest osteoinductive potential among all BMP family members [58] and, therefore, has been incorporated with orthopedic implants for increasing bone growth after implantation. To reduce infection and inflammation, well-known antibiotics such as penicillin and gentamicin (an aminoglycoside antibiotic particularly useful for treating gram-negative infection) have also been loaded to orthopedic implants.

Nanotechnology has enabled better controls on delivery efficiency and release kinetics of loaded drugs, leading to improved bone implant efficacy [57]. For instance, immobilizing a segment of BMP-2 (the knuckle epitope, CKIPKASSVPTELSAISTLYL) into anodized Ti nanotubes (approximately 70 nm in diameter and 200 nm in depth) can further promote osteoblast adhesion (note that anodized Ti itself can increase osteoblast adhesion compared to conventional Ti, as mentioned before) [59]. Gentamicin loaded anodized Ti nanotubes revealed a prolonged continuous release of the drug, leading to decrease in *S. epidermidis* colonization and enhancement of osteoblast preosteoblastic cell (MC3T3-E1) differentiation on the metal surface [35] (Figure 2.10). Another study stored drugs into chemically functionalized nanotubes of anodized Ti to increase the drug-loading efficiency and prolong drug release by a silanization process (unpublished results by Yao C and Webster TJ). This study used penicillin G as a drug model and revealed that drug release inhibited infection by significantly reducing bacterial density (*S. epidermidis*) on the implant. The same researchers also reported that immersing anodized Ti into a mixture of simulated body fluid and penicillin-based drug could create a drug-loaded calcium phosphate coating on the Ti implant surface that prolonged drug release for up to 3 weeks [60].

In summary, studies and investigations conducted so far indicate there is an opportunity of using metallic nanosurfaces for controlling drug release and improving bone implant efficacy. It has also been demonstrated that nanotechnology can achieve high loading efficiency and prolonged release of drugs in metallic orthopedic implants that are currently used.

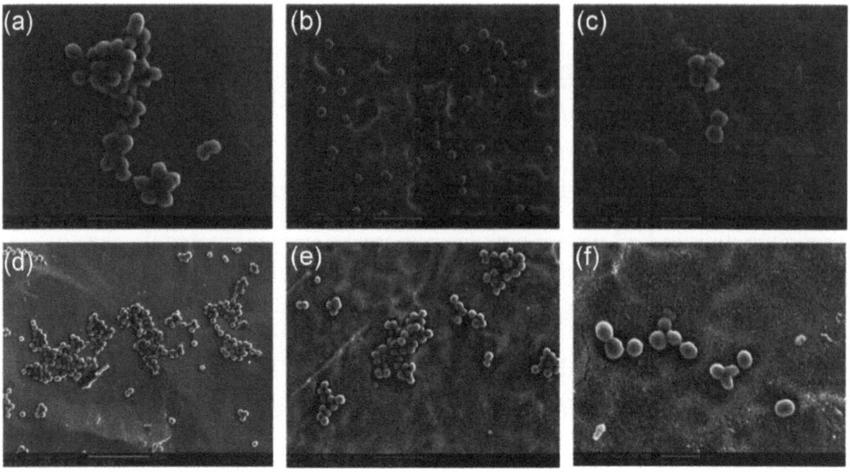

Figure 2.10 SEM images of bacterial colonies after 1 h on (a) Ti without antibiotic, (b) anodized Ti nanotubes without antibiotic, (c) nanotubes filled with gentamicin; and after 4 h on (d) Ti without antibiotic, (e) anodized Ti nanotubes without antibiotic, and (f) nanotubes filled with gentamicin.
Modified with permission from Ref. [35].

2.3.3 Metallic nanoparticles for sensing and detection

Recent developments of nanotechnology have extended the scope of biomaterials to "smart biomaterials" that can simultaneously incorporate intelligent functionalities such as disease sensing, diagnosis, imaging, and treatment [57]. For orthopedic applications, *in situ* sensing of the bone tissue growth or infection/inflammation responses surrounding orthopedic implants is of great interest [61]. Although the relevant studies are still in their early stages and mainly focus on evaluating sensing, detecting and biocompatibility properties of materials, this is a highly active and rapid developing research area. A few examples of metallic nanoparticles possessing great potential for sensing and detection purposes are briefly summarized below.

Au and Ag nanoparticles have been extensively studied as better medical imaging contrast agents and drug carriers with detection capability to target specific tissues (e.g., cancer sites) *in vivo* [11,62,63]. Both Au and Ag have outstanding plasmon resonance effects that can be employed for labeling and tracking cells and tissues with high precision. Besides, upon binding of proteins and/or dyes to Au and Ag nanoparticles, changes in both the optical intensity and the wavelength of the particles are observed and thus can be used to detect specific proteins [62]. Au and Ag have also been employed as glucose sensing platforms after the modification of their surfaces with dextran and boronic acid groups, respectively [64]. By using surface-modified Au and Ag, simple optical density and luminescence emission measurements were able to detect glucose levels with tunable sensitivity at physiologically compatible observation wavelengths.

Magnetic nanoparticles like FeCo and FePt have been explored for biodetection applications based on their magnetoresistive properties [65,66]. Such nanoparticles are excellent magnetic markers to identify biomolecules and measure concentrations of biomolecules with high sensitivity. For example, a highly sensitive giant-magnetoresistive chip and FeCo nanoparticles were devised and revealed the ability to linearly detect 600–4500 copies of streptavidin. Under normal conditions, this system could also detect human IL-6 with a sensitivity 13 times higher than that of standard enzyme-linked immunosorbent assay techniques [67].

2.4 Nanostructured metallic implants with superior mechanical properties

There are extensive studies centering on the fabrication of nanostructured metals in order to improve their mechanical properties. Since mechanical performance of orthopedic implant is critical to its applications, liability and lifetime, superior mechanical properties are always wanted. Depending on clinical settings, the wanted properties include, but are not limited to, enhanced mechanical strengths, toughness, ductility, wear resistance, corrosion resistance, and special characteristics such as superplasticity and shape-memory effect. Due to space limitations, only the typical aspects and examples of implant mechanical properties enhanced by nanotechnology are introduced here.

Both strength and ductility of orthopedic metals can be improved by creating nanostructures or nanophases in the materials. For instance, 316L austenitic stainless steel that is frequently used in orthopedics could be processed by means of dynamic plastic deformation with high strain rates and thermal annealing to obtain microsized grains embedded with nanograins and nanometer-thick twinning structures [68]. This nanostructured stainless steel exhibits an enhanced strength–ductility synergy and much higher work-hardening rates than the as-deformed conventional stainless steel. A combination of 1.0GPa tensile strength (conventional 316L austenitic stainless steel is around 0.3GPa) with an elongation-to-failure strain of ~27% was achieved in the nanostructured stainless steel. Also, when the strain was below 24%, the work-hardening rates of the nanostructured stainless steel were even higher than that of the conventional, coarse-grained stainless steel.

Improved corrosion resistances are also reported in nanotechnology-enhanced metals. The effect of nanocrystallization on the corrosion behavior of 309 stainless steel was investigated in sodium sulfate (Na_2SO_4, pH=2) solution [69]. Electrochemical evaluations revealed that the nanocrystalline 309 stainless steel had a smaller corrosion current density than conventional steel, indicating that corrosion resistance of the stainless steel was improved by nanocrystallization of the stainless steel. Nanocrystalline Ni–P alloy coatings were also developed via electrode-less deposition to protect bulk materials from corrosion, and this type of corrosion protective coating is among the earliest industrial applications of nanocrystalline metals [70]. For antiwear purposes, metallic coatings produced by nanotechnology have received lots of attention too. Ni–SiC nanocomposite (5% SiC added to Ni) coatings were developed and investigated for wear-resistant applications [71]. The results showed that Ni–SiC nanocomposite coating had 30% higher microhardness and 54% lower wear rate than Ni coating, probably due to grain fining and dispersive strengthening effects of the SiC nanoparticles added to Ni matrix. In addition, the Ni–SiC nanocomposite coating showed smaller passive current density than Ni coating in electrochemical corrosion test, suggesting that the Ni–SiC nanocomposite coating had better corrosion resistance than Ni coating.

In addition, alloys possessing shape-memory effects are of great interest for biomedical applications. Shape memory effect occurs when a material returns to, or "remembers," its original (or cold-forged) shape after being heated. The family of shape-memory alloys (SMAs) includes Ni–Ti alloys and copper-base alloys (e.g., Cu–Zn–Al and Cu–Al–Ni). NiTi SMAs are used in surgical implants and biomedical devices such as orthodontic wires, spine correction rods, bone fraction fixation plate and staples [72]. Nanotechnology-enhanced orthopedic SMAs is becoming a research area with rapidly growing interest, and nanophase SMAs with enhanced mechanical strength, corrosion or wear resistance, and promoted cell interactions have been developed [72–74]. For example, in order to enhance the surface wear resistance, orthopedic NiTi alloy was subjected to surface mechanical attrition treatment (SMAT) and nanocrystalline and partial amorphous structures were fabricated on the metal surface. It was found that hardness in the surface layer was notably improved and the corrosion resistance of SMAT-treated NiTi was significantly enhanced compared to bare NiTi with coarse grains [72]. A recent study applied Zr ion implantation to NiTi alloy and formed a Ni-depleted nanocomposite layer on the NiTi alloy surface [73]. The nanocomposite layer resulted in increased corrosion resistance, reduced Ni ion leaching, and promoted endothelial cell proliferation on the alloy surfaces as compared to conventional NiTi alloy.

2.5 Commercialization status of nanostructured metallic implants

Although nanotechnology-enhanced metals and alloys have demonstrated extraordinary properties and promises for orthopedic applications in the laboratory, the commercialization and approval of the aforementioned nanomaterials for clinical use still need time. This long wait is largely because the extensive evaluations on safety of these nanomaterials have yet to be completed and the long-term efficacies of such nanomaterials are still not fully known. Given that most of the nanomaterials were invented in the past two decades, this challenge will probably remain for an extended time. Nevertheless, nanomedicine is a fast growing entrepreneurial area where a good number of companies are involved in research and development, commercialization, and even publicization of nanomaterials for medical applications. For example, most of the major players and established multinational companies in biomedical field have internal research programs on drug delivery that are based on formulations or dispersions containing nanomaterial components [62].

For orthopedic implant applications, a number of joint ventures and multinationals have recognized the potential and importance of nanotechnology and are now transferring the technologies of modifying metals at nanoscale from bench to bed. For example, a joint venture of Manhattan Scientifics and BASIC Dental has exclusively licensed a patented nanostructuring technology from the Los Alamos National Laboratory to make it available for Ti orthopedic implants. The venture's first product of dental implant has obtained the approval of the US Food and Drug Administration (FDA) to manufacture and market a type of nanostructured Ti implant, which made its US debut in 2010 [75]. A group of similar techniques for the fabrication of nanostructured metals and alloys based on techniques have also entered the commercialization pathway in the past decade [76]. As mentioned in section 2.1.4, SPD techniques are of great value in producing bulk nanometals and alloys for orthopedic implants. In addition, nanostructured steels containing nanoscale bainite and austenite has been fully commercialized and can be now purchased in large quantity (e.g., in tons) [77]. These steels possess a strength as high as 2.5 GPa and, meanwhile, excellent ductility, toughness, fatigue resistance, and wear resistance. Although these steels have not been reported for orthopedic use, the variants of such steels are likely to be extended to orthopedic implant applications due to their superior mechanical properties and commercial availability.

There are also some start-ups or joint ventures dedicated to commercializing the technologies that create surface nanopatterns or nanoroughness on currently used orthopedic implants. For instance, companies like Nanovis, Inc. (Indiana, US), have developed or patented a variety of surface patterning or roughening techniques for orthopedic implants and plan to release new products in the next few years. However, most of their current efforts are in preclinical trials and/or obtaining the US FDA approval.

2.6 Summary

Nanotechnology-enhanced metals and alloys are promising orthopedic materials for controlling protein adsorption and cell functions at the bone–implant interface, expanding implant functionalities, and improving mechanical properties of orthopedic implants. Nanostructured metals and alloys, including both particulate and bulk forms, can be fabricated via a variety of top-down or bottom-up techniques. The extraordinary *in vitro* and *in vivo* performances of nanostructured metals and alloys allow next-generation orthopedic implants to better integrate with human bone tissue and serve for longer times. However, safety concerns on nanostructured orthopedic materials remains and should be thoroughly investigated. Nevertheless, nanostructured metals and alloys are of substantial clinical and commercial values and a number of companies have reached different stages of commercialization.

References

[1] Siegel RW. Creating nanophase materials. Sci Am 1996;275:74–9.
[2] Choi C, Dong X, Kim B. Characterization of Fe and Co nanoparticles synthesized by chemical vapor condensation. Scr Mater 2001;44:2225–9.

[3] Singh J, Wolfe D. Review: nano and macro-structured component fabrication by electron beam-physical vapor deposition (EB-PVD). J Mater Sci 2005;40:1–26.

[4] Bahmanpour H, Kajbafvala A, Maneshian MH, Zargar HR, Youssef K. Bulk nanostructured metals and alloys: processing, structure, and thermal stability. J Nanomater 2012;2012:2.

[5] Attar A, Halali M, Sobhani M, Ghandehari RT. Synthesis of titanium nano-particles via chemical vapor condensation processing. J Alloys Compd 2011;509:5825–8.

[6] Wang Z, Zhang Z, Choi C, Kim B. Structure and magnetic properties of Fe (C) and Co (C) nanocapsules prepared by chemical vapor condensation. J Alloys Compd 2003;361:289–93.

[7] Dong X, Choi C, Kim B. Structural and magnetic characterization of Fe nanoparticles synthesized by chemical vapor condensation process. J Appl Phys 2002;92:5380–5.

[8] Koch C. Synthesis of nanostructured materials by mechanical milling: problems and opportunities. Nanostruct Mater 1997;9:13–22.

[9] Fecht H, Hellstern E, Fu Z, Johnson W. Nanocrystalline metals prepared by high-energy ball milling. Metall Trans A 1990;21:2333–7.

[10] Brown TL, Swaminathan S, Chandrasekar S, Compton WD, King AH, Trumble KP. Low-cost manufacturing process for nanostructured metals and alloys. J Mater Res 2002;17:2484–8.

[11] Skrabalak SE, Chen J, Sun Y, Lu X, Au L, Cobley CM, et al. Gold nanocages: synthesis, properties, and applications. Acc Chem Res 2008;41:1587–95.

[12] Xia Y, Xiong Y, Lim B, Skrabalak SE. Shape-controlled synthesis of metal nanocrystals: simple chemistry meets complex physics? Angew Chem Int Ed 2009;48:60–103.

[13] Segal V, Reznikov V, Drobyshevskiy A, Kopylov V. Plastic metal working by simple shear. Russ Metall 1981;1:99–104.

[14] Iwahashi Y, Wang J, Horita Z, Nemoto M, Langdon TG. Principle of equal-channel angular pressing for the processing of ultra-fine grained materials. Scr Mater 1996;35:143–6.

[15] Le Guehennec L, Soueidan A, Layrolle P, Amouriq Y. Surface treatments of titanium dental implants for rapid osseointegration. Dent Mater 2007;23:844–54.

[16] Yao C, Webster TJ. Anodization: a promising nano-modification technique of titanium implants for orthopedic applications. J Nanosci Nanotechnol 2006;6:9–10.

[17] Allam NK, Feng XJ, Grimes CA. Self-assembled fabrication of vertically oriented Ta_2O_5 nanotube arrays, and membranes thereof, by one-step tantalum anodization. Chem Mater 2008;20:6477–81.

[18] Yang L, Zhang LJ, Webster TJ. Nanobiomaterials: state of the art and future trends. Adv Eng Mater 2011;13:B197–217.

[19] Karthikeyan J, Berndt C, Tikkanen J, Reddy S, Herman H. Plasma spray synthesis of nanomaterial powders and deposits. Mater Sci Eng A 1997;238:275–86.

[20] Boulos MI, Fauchais P, Vardelle A, Pfender E. Fundamentals of plasma particle momentum and heat transfer. In: Suryanarayan R, editor. Plasma spraying: theory and applications. Singapore: World Scientific; 1993. p. 3–60.

[21] Laha T, Agarwal A, McKechnie T, Rea K, Seal S. Synthesis of bulk nanostructured aluminum alloy component through vacuum plasma spray technique. Acta Mater 2005;53:5429–38.

[22] Cui C, Liu H, Li Y, Sun J, Wang R, Liu S, et al. Fabrication and biocompatibility of nano-TiO_2/titanium alloys biomaterials. Mater Lett 2005;59:3144–8.

[23] Zhu Y-C, Yukimura K, Ding C-X, Zhang P-Y. Tribological properties of nanostructured and conventional WC–Co coatings deposited by plasma spraying. Thin Solid Films 2001;388:277–82.

[24] Li B, Liu X, Meng F, Chang J, Ding C. Preparation and antibacterial properties of plasma sprayed nano-titania/silver coatings. Mater Chem Phys 2009;118:99–104.

[25] Poorna Chander K, Vashista M, Sabiruddin K, Paul S, Bandyopadhyay P. Effects of grit blasting on surface properties of steel substrates. Mater Des 2009;30:2895–902.

[26] Nayak BK, Gupta MC. Self-organized micro/nano structures in metal surfaces by ultrafast laser irradiation. Opt Lasers Eng 2010;48:940–9.

[27] Long M, Rack H. Titanium alloys in total joint replacement—a materials science perspective. Biomaterials 1998;19:1621–39.

[28] Bannon B, Mild E. Titanium alloys for biomaterial application: an overview, titanium alloys in surgical implants. ASTM STP 1983;796:7–15.

[29] Geetha M, Singh A, Asokamani R, Gogia A. Ti based biomaterials, the ultimate choice for orthopaedic implants–a review. Prog Mater Sci 2009;54:397–425.

[30] Yang L, Webster T. Biological responses to and toxicity of nanoscale implant materials. In: Eliaz N, editor. Degradation of implant materials. New York: Springer; 2013. p. 481–508.

[31] Boyan B, Lossdorfer S, Wang L, Zhao G, Lohmann C, Cochran D, et al. Osteoblasts generate an osteogenic microenvironment when grown on surfaces with rough microtopographies. Eur Cell Mater 2003;6:22–7.

[32] Christenson EM, Anseth KS, van den Beucken JJ, Chan CK, Ercan B, Jansen JA, et al. Nanobiomaterial applications in orthopedics. J Orthop Res 2007;25:11–22.

[33] Yao C, Perla V, McKenzie JL, Slamovich EB, Webster TJ. Anodized Ti and Ti6Al4V possessing nanometer surface features enhances osteoblast adhesion. J Biomed Nanotechnol 2005;1:68–73.

[34] Webster TJ, Ejiofor JU. Increased osteoblast adhesion on nanophase metals: Ti, Ti6Al4V, and CoCrMo. Biomaterials 2004;25:4731–9.

[35] Popat KC, Eltgroth M, LaTempa TJ, Grimes CA, Desai TA. Decreased Staphylococcus epidermis adhesion and increased osteoblast functionality on antibiotic-loaded titania nanotubes. Biomaterials 2007;28:4880–8.

[36] Huang H-H, Pan S-J, Lai Y-L, Lee T-H, Chen C-C, Lu F-H. Osteoblast-like cell initial adhesion onto a network-structured titanium oxide layer. Scr Mater 2004;51:1017–21.

[37] Zhu X, Chen J, Scheideler L, Reichl R, Geis-Gerstorfer J. Effects of topography and composition of titanium surface oxides on osteoblast responses. Biomaterials 2004;25:4087–103.

[38] Das K, Bose S, Bandyopadhyay A. Surface modifications and cell–materials interactions with anodized Ti. Acta Biomater 2007;3:573–85.

[39] Tambasco de Oliveira P, Nanci A. Nanotexturing of titanium-based surfaces upregulates expression of bone sialoprotein and osteopontin by cultured osteogenic cells. Biomaterials 2004;25:403–13.

[40] Cohen A, Liu-Synder P, Storey D, Webster TJ. Decreased fibroblast and increased osteoblast functions on ionic plasma deposited nanostructured Ti coatings. Nanoscale Res Lett 2007;2:385–90.

[41] Ward BC, Webster TJ. The effect of nanotopography on calcium and phosphorus deposition on metallic materials in vitro. Biomaterials 2006;27:3064–74.

[42] Yao C, Slamovich EB, Webster TJ. Increased osteoblast adhesion on nano-rough anodized titanium and CoCrMo. In: NSTI nanotechnology conference and trade show—NSTI nanotech, technical proceedings; 2006. p. 119–22.

[43] Nakanishi Y, Miura H, Tokunaga K, Hidehiko H, Mizuta H, Iwamoto Y. Nano-level surface texturing on Co–Cr–Mo alloy inhibits macrophage activation in joint prostheses. Bone Joint J Orthop Proc Suppl 2013;95:278.

[44] Papageorgiou I, Brown C, Schins R, Singh S, Newson R, Davis S, et al. The effect of nano- and micron-sized particles of cobalt–chromium alloy on human fibroblasts in vitro. Biomaterials 2007;28:2946–58.

[45] Tran P, Webster TJ. Enhanced osteoblast adhesion on nanostructured selenium compacts for anti-cancer orthopedic applications. Int J Nanomed 2008;3:391.

[46] Tran PA, Sarin L, Hurt RH, Webster TJ. Opportunities for nanotechnology-enabled bioactive bone implants. J Mater Chem 2009;19:2653–9.

[47] Navarro-Alarcon M, López-Martínez MC. Essentiality of selenium in the human body: relationship with different diseases. Sci Total Environ 2000;249:347–71.

[48] Kopeikin V, Valueva S, Kipper A, Borovikova L, Filippov A. Synthesis of selenium nanoparticles in aqueous solutions of poly (vinylpyrrolidone) and morphological characteristics of the related nanocomposites. Polym Sci Ser A: Chem Phys 2003;45:374–9.

[49] Puckett SD, Taylor E, Raimondo T, Webster TJ. The relationship between the nanostructure of titanium surfaces and bacterial attachment. Biomaterials 2010;31:706–13.

[50] Truong VK, Lapovok R, Estrin YS, Rundell S, Wang JY, Fluke CJ, et al. The influence of nano-scale surface roughness on bacterial adhesion to ultrafine-grained titanium. Biomaterials 2010;31:3674–83.

[51] Ercan B, Taylor E, Alpaslan E, Webster TJ. Diameter of titanium nanotubes influences anti-bacterial efficacy. Nanotechnology 2011;22:295102.

[52] Zhao L, Wang H, Huo K, Cui L, Zhang W, Ni H, et al. Antibacterial nano-structured titania coating incorporated with silver nanoparticles. Biomaterials 2011;32:5706–16.

[53] Ionita D, Ungureanu C, Demetrescu I. Electrochemical and antibacterial performance of CoCrMo alloy coated with hydroxyapatite or silver nanoparticles. J Mater Eng Perform 2013;22:1–8.

[54] Tran PA, Webster TJ. Selenium nanoparticles inhibit *Staphylococcus aureus* growth. Int J Nanomed 2011;6:1553.

[55] Wang Q, Webster TJ. Nanostructured selenium for preventing biofilm formation on polycarbonate medical devices. J Biomed Mater Res A 2012;100:3205–10.

[56] Holinka J, Pilz M, Kubista B, Presterl E, Windhager R. Effects of selenium coating of orthopaedic implant surfaces on bacterial adherence and osteoblastic cell growth. Bone Joint J 2013;95:678–82.

[57] Yang L, Webster TJ. Nanotechnology controlled drug delivery for treating bone diseases. Expert Opin Drug Del 2009;6:851–64.

[58] Laub M, Jennissen H, Seul T, Schmachtenberg E. Molecular modelling of bone morphogenetic protein-2 (BMP-2) by 3D-rapid prototyping. Materialwissenschaft Und Werkstofftechnik 2001;32:926–30.

[59] Balasundaram G, Yao C, Webster TJ. TiO_2 nanotubes functionalized with regions of bone morphogenetic protein-2 increases osteoblast adhesion. J Biomed Mater Res A 2008;84:447–53.

[60] Yao C, Webster TJ. Prolonged antibiotic delivery from anodized nanotubular titanium using a co-precipitation drug loading method. J Biomed Mater Res B Appl Biomater 2009;91:587–95.

[61] Webster TJ, editor. Nanotechnology enabled in situ sensors for monitoring health. New York: Springer; 2011. 235 pp.

[62] Sironmani A, Daniel K. Silver nanoparticles-universal multifunctional nanoparticles for bio sensing, imaging for diagnostics and targeted drug delivery for therapeutic applications. In: Kapetanovic DI, editor. Drug discovery and development—present and future. Rijeka, Croatia: InTech; 2011. p. 463–88.

[63] Lee K-S, El-Sayed MA. Gold and silver nanoparticles in sensing and imaging: sensitivity of plasmon response to size, shape, and metal composition. J Phys Chem B 2006;110:19220–5.
[64] Aslan K, Zhang J, Lakowicz JR, Geddes CD. Saccharide sensing using gold and silver nanoparticles—a review. J Fluoresc 2004;14:391–400.
[65] Reiss G, Brueckl H, Huetten A, Schotter J, Brzeska M, Panhorst M, et al. Magnetoresistive sensors and magnetic nanoparticles for biotechnology. J Mater Res 2005;20:3294.
[66] Koh I, Josephson L. Magnetic nanoparticle sensors. Sensors 2009;9:8130–45.
[67] Srinivasan B, Li Y, Jing Y, Xu Y, Yao X, Xing C, et al. A detection system based on giant magnetoresistive sensors and high-moment magnetic nanoparticles demonstrates zeptomole sensitivity: potential for personalized medicine. Angew Chem Int Ed 2009;48:2764–7.
[68] Yan F, Liu G, Tao N, Lu K. Strength and ductility of 316L austenitic stainless steel strengthened by nano-scale twin bundles. Acta Mater 2012;60:1059–71.
[69] Ye W, Li Y, Wang F. The improvement of the corrosion resistance of 309 stainless steel in the transpassive region by nano-crystallization. Electrochim Acta 2009;54:1339–49.
[70] Crobu M, Scorciapino A, Elsener B, Rossi A. The corrosion resistance of electroless deposited nano-crystalline Ni–P alloys. Electrochim Acta 2008;53:3364–70.
[71] Vaezi M, Sadrnezhaad S, Nikzad L. Electrodeposition of Ni–SiC nano-composite coatings and evaluation of wear and corrosion resistance and electroplating characteristics. Colloids Surf A Physicochem Eng Asp 2008;315:176–82.
[72] Hu T, Xin Y, Wu S, Chu C, Lu J, Guan L, et al. Corrosion behavior on orthopedic NiTi alloy with nanocrystalline/amorphous surface. Mater Chem Phys 2011;126:102–7.
[73] Zhao T, Li Y, Xia Y, Venkatraman SS, Xiang Y, Zhao X. Formation of a nano-pattering NiTi surface with Ni-depleted superficial layer to promote corrosion resistance and endothelial cell–material interaction. J Mater Sci Mater Med 2013;24:105–14.
[74] Wu S, Liu X, Chu PK, Hu T, Yeung KWK, Chung JC. Fabrication and surface modification of porous nano-structured NiTi orthopedic scaffolds for bone implants. In: MRS proceedings. Cambridge University Press; 2009.
[75] Lowe TC. Nano-structured metals. Today's Medical Development 2011;(August). http://www.onlinetmd.com/Magazine/ [accessed 25.09.13].
[76] Lowe TC. Metals and alloys nanostructured by severe plastic deformation: commercialization pathways. JOM 2006;58:28–32.
[77] Bhadeshia HKDH. The first bulk nanostructured metal. Sci Technol Adv Mater 2013;14:014202.

Orthopedic nanoceramics

3

The field of nanoceramics for biomedical applications is one of the fastest emerging and developing arenas drawing much attention over the past decade. Nanoceramics usually indicate nanophase ceramics whose feature (i.e., grain or particle) sizes are within the nanoscale range (1–100 nm), including at least four different forms: particulate (zero-dimensional, 0-D), whisker or wire (1-dimensional, 1-D), coating or thin film (2-dimensional, 2-D), and scaffold (3-dimensional, 3-D). For orthopedic applications, attractive characteristics of nanophase ceramics (including their size, structural advantages, unique physical and chemical properties, and ease of modification) make them competent platforms for structural support, bone regeneration, drug distribution, and other orthopedic uses. Due to their geometric and dimensional differences, nanophase ceramics have been prepared via different fabrication methods and studied for different orthopedic applications. Generally, ceramic nanoparticles and nanowhiskers are frequently combined with other materials like polymers to produce nanostructured composite bone scaffolds and drug carriers. In contrast, ceramic nanocoatings are usually used for the improvements of biocompatibility and wear resistance of bone implants, whereas nanoscaffolds are mainly used for direct substitution of defective bone. In this chapter, fabrication and application of each type of nanoceramic will be introduced and reviewed. As a footnote, this chapter mainly focuses on ceramic materials, while the hybrid or composite nanomaterial systems that consist of both ceramics and other materials like polymers will be introduced in Chapter 4.

3.1 Fabrication of nanoceramics

Owing to their architectural difference, fabrication methods of nanoscale ceramic particles, coatings, and scaffolds vary dramatically. This section introduces the fabrication methods of each category individually. It is worth noting that, in contrast to metals and polymers, fabrication of ceramic materials usually includes heat treatments such as calcination or sintering at elevated temperatures. In fact, calcination or sintering is of great importance to nanoceramic manufacture and plays a crucial role in the material and biological properties of nanoceramic.

3.1.1 Synthesis of ceramic nanoparticles

There are a great number of synthesis methods for various ceramic nanoparticles but they can be essentially divided into two categories: top-down and bottom-up methods. As introduced in Chapter 1, both categories have pros and cons in producing nanomaterials. Because medical applications usually require materials to have high

Nanotechnology-Enhanced Orthopedic Materials. http://dx.doi.org/10.1016/B978-0-85709-844-3.00003-3

purity, high biological safety and biocompatibility, bottom-up synthesis approaches are advantageous over top-down methods due to their better controls in the quality, purity, structure, and property of nanomaterials. Hence, bottom-up methods, especially wet-chemical synthesis routes, are introduced here.

Sol–gel and precipitation methods are simple and commonly used wet-chemical synthesis methods of ceramic nanoparticles such as calcium phosphates, iron oxides, silica, titanium oxides, and zinc oxides. Basically, the sol–gel method uses inorganic precursors (i.e., meal salts or organometallic molecules) that react in aqueous environment and subsequently form integrated network (gel). For example, metal oxide nanoparticles are often synthesized via the hydrolysis and condensation reactions of metal alkoxides:

Hydrolysis:

$$\mathrm{M(OR)}_x + \mathrm{H_2O} \rightarrow \mathrm{M(OR)}_{x-1}\mathrm{OH} + \mathrm{ROH} \cdots \rightarrow \mathrm{M(OH)}_x + x\mathrm{ROH}$$

Condensation:

$$2\mathrm{M(OR)}_{x-1}\mathrm{OH} \rightarrow \mathrm{M(OR)}_{x-1}\mathrm{OM(OR)}_{x-1} + \mathrm{H_2O}$$

$$\mathrm{M(OR)}_{x-1}\mathrm{OH} + \mathrm{M(OR)}_{x-1}\mathrm{OR} \rightarrow \mathrm{M(OR)}_{x-1}\mathrm{OM(OR)}_{x-1} + \mathrm{ROH}$$

where M is the metal and R the alkyl group [1,2]. The gel resulted from the condensation reactions is usually filtered, washed, and rinsed-off residual ions, and then calcined at high temperatures (e.g., several hundred degree centigrade) to generate well-crystallized powders or particles [3,4]. A myriad of metal oxides or metal oxide composites can be produced by this approach, including iron oxides, tin oxides, silicon oxides, zinc oxides, and titanium oxides. In addition to metal oxides, sol–gel methods can also be employed to prepare phosphate nanoparticles such as calcium phosphates [5] and $LiFePO_4$ [6]. More importantly, sol–gel methods have been developed for fabricating complex ceramic nanostructures such as porous [7] and core–shell [8] structures, and for modifying nanoceramic surfaces [8,9].

Precipitation (or coprecipitation) method is also a simple and efficient wet-chemical route for preparing ceramic nanoparticles. A carefully developed precipitation method that optimizes processing parameters such as reactants and their concentrations, pH, temperature, and calcination conditions can produce a massive and reproducible quantity of ceramic nanoparticles with high purity and crystallinity [10]. For example, one of the most important ceramics in orthopedic applications, nanocrystalline hydroxyapatite ($Ca_{10}(PO_4)_6(OH)_2$, HA), can be produced in large quantity through the aqueous reaction:

$$10\mathrm{Ca(NO_3)_2}\cdot 4\mathrm{H_2O} + 6\mathrm{(NH_4)_2HPO_4} + 8\mathrm{NH_4OH} \rightarrow \mathrm{Ca_{10}(PO_4)_6(OH)_2}\downarrow + 20\mathrm{NH_4NO_3} + 6\mathrm{H_2O}$$

The advantage of this precipitation reaction is that NO_3^- and NH_4^+ ions are claimed not to incorporate into HA lattice, leading to highly pure crystalline product [11].

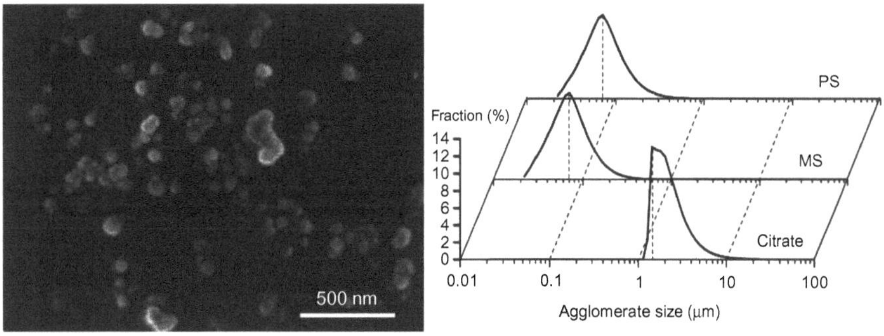

(a) (b)

Figure 3.1 (a) SEM image of HA nanocrystals synthesized with gelatinized starch matrix; (b) dynamic light scattering (DLS) size distributions of HA agglomerates synthesized with gelatinized starch matrices (PS, potato starch; MS, modified potato starch) and citrate, *x* axis is agglomerate size and *y* axis is particle fraction.

Similar to the sol–gel method, the precipitates from the aqueous reaction are usually washed, dried, and calcined to yield well-crystallized crystals.

However, both the sol–gel and precipitation methods frequently encounter the agglomeration problem. Ceramics particles produced by these methods are usually heavily agglomerated, causing agglomerates of sizes up to a few hundred microns. Primary agglomeration of primitive crystallites occurs during the reaction, precipitation or aging processes [12,13], while secondary yet elevated agglomeration often occurs in subsequent isothermal treatment (e.g., hydrothermal treatment or calcination) at temperatures ranging from a few hundred to over one thousand degree centigrade with holding times varying from several to tens of hours [13–16]. To relieve the agglomeration, surfactants such as sodium citrate, castor oil, silicone coupling agents, and other commercially available dispersants are always added to the liquid during wet-synthesis of nanoparticles [17,18]. Recently, a new method used gelatinized starch matrix instead of surfactant to produce dispersed nanoparticles in large quantity [19]. Figure 3.1 shows the nonagglomerated HA nanocrystals synthesized in assistance with the gelatinized starch matrix and the medium size (d_{50}) of the HA nanocrystals was about 65 nm.

Several other synthesis methods such as hydrolysis [20], pyrolysis [21,22], hydrothermal [10,23], and free-drying [24] methods are often used to fabricate ceramic nanoparticles, including calcium phosphates and carbonates, metal oxides, as well as nonoxides such as metal sulfates. Due to space limitations, these methods are not expanded here but it is important to note that the versatility of these methods provides rich opportunities to manufacture, modify, and functionalize complex nanoparticles or other nanoarchitectures.

3.1.2 Fabrication of ceramic nanocoatings

Ceramic nanocoatings are generally defined as 2-D ceramic thin films that have crystalline or surface feature (e.g., grain size or thickness) sizes below a few hundred nanometers. Depending on material composition and structure, well-established fabrication methods of ceramic nanocoatings include wet-chemical routes, plasma and/or spray-coating techniques, and solid-state reactions. For biomedical materials, wet-chemical coating approaches usually refer to sol–gel processes that employ hydrolysis and condensation reactions of precursor molecules to cover a material with a layer of gel [2]. The gel can be transferred or deposited on the material by coating techniques such as dip coating, spin coating, electrospinning, and so on, and then the gel coating is usually treated by heat to transform into solid and uniform coating, as illustrated in Figure 3.2. Specifically for orthopedic applications, sol–gel process is frequently used to coat calcium phosphates (e.g., HA) on orthopedic implant surfaces [25,26]. For example, $Ca(NO_3)_2{\cdot}4H_2O$ and $PO(CH_3)_3$ precursors were used to prepare a gel and the gel was dip-coated on Ti implant surface and then calcined at 400 °C to obtain a nanocrystalline HA coating on Ti [27]. The sol–gel process using alkoxide precursors to prepare HA has been applied to generate crack-free nanocrystalline HA coatings on a variety of substrates including Al_2O_3, Vycor glass, ZrO_2, Ti6Al4V alloy, and single crystal MgO [28]. Similarly, precipitation method and hydrothermal method have also been adapted to fabricate ceramic nanocoatings on various substrate materials [27].

In the past few decades, plasma spray-coating techniques have been developed to cover orthopedic implants with protective and/or bioactive coatings. As introduced in Chapter 1, the plasma spray-coating method employs high temperature plasma jet to melt and spray a feedstock material onto a substrate to form a coating. The feedstock materials for plasma spray can be in the forms of solid, liquid or suspension [29,30]. For the fabrication of nanocoating on orthopedic implants, the commonly used solid

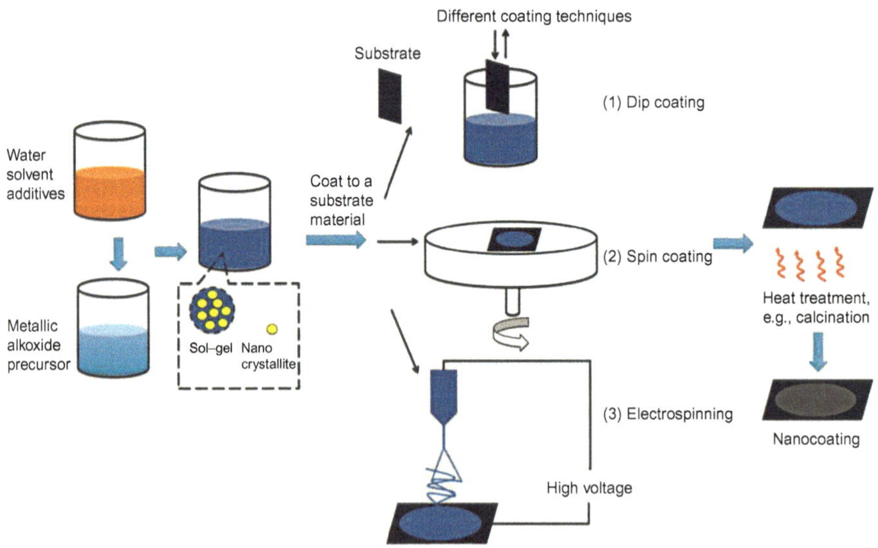

Figure 3.2 Illustration of sol–gel process of preparing nanocoating via three different coating techniques: dip coating, spin coating, and electrospinning.

feedstock materials include micron-sized HA or calcium phosphate powders while the regular liquid feedstock materials are various types of calcium phosphate suspensions formed from precursors such as calcium nitrate and diammonium nitrate [31,32]. Besides HA, Si-incorporated HA, antibacterial Ag-contained HA, and TiO_2–HA functionally graded coatings have been successfully deposited on metallic implant surfaces by plasma spray techniques [31].

Nanocoatings of TiO_2, HA and its derivatives, and bioactive glass can also be coated on metallic implants by electrophoretic deposition [33–35]. In a typical electrophoretic deposition, direct current is employed to charge the nanometer or submicron powder particles suspended in a liquid medium and drive them to deposit onto a conductive substrate (e.g., metallic implant surface) of opposite charge. This technique is facile for fabricating homogeneous and dense ceramic nanocoatings and, thus, has been applied to coat a variety of metallic implants including Ti and its alloys, stainless steels, and Mg and its alloys.

3.1.3 Fabrication of ceramic nanoscaffolds

Ceramic nanoscaffolds have drawn a lot of attention for bone tissue engineering applications, largely because they can provide both structural (or mechanical) and functional supports for bone ingrowth. Since bone ingrowth generally requires interconnected pores or channels of sizes more than 100 μm, ceramic nanoscaffolds for orthopedic applications are usually 3-D porous structures. The fabrication processes of porous nano-scaffold are pertinent to their efficacy in supporting the cell or tissue activities and, thus, have been extensively studied. Based on the techniques of creating pores, there are at least five types of fabrication techniques.

Removable template techniques. These techniques use thermally removable or soluble templates (also known as porogens) to generate pores in molded ceramics after sintering or dissolving. Common templates or porogens include polymeric beads or sponges (e.g., polyurethane, polyethylene glycol) and salt crystals (e.g., NaCl, KCl) that can either thermally decompose or leach into specific solvents in the subsequent treatments [36,37]. Figure 3.3a shows the porous β-TCP scaffold spark-plasma-sintered from nanocrystalline powder using 350–650 μm polyethylene glycol particles as porogens [38]. The advantages of removable template techniques include adjustable porous structure and highly controllable porosity or pore interconnectivity.

Gas-foaming techniques. Gas foaming provides another facile way to fabricate porous ceramic nanoscaffolds with controllable porous structure and porosity. The method is based on the generation of foam from an aqueous suspension of ceramic powder and subsequent stabilization of the structure by *in situ* polymerization of organic monomers or solidification of the suspension. The gas-foaming method is both time and cost efficient for preparing 3-D porous structures with high porosity (usually >40%), micrometer pore sizes (usually a few to several hundred microns), and large-size structures that are suitable for bone substitution applications (e.g., up to decimeters in their external dimensions). Figure 3.3b shows a porous HA scaffold prepared by gas-foaming technique, which had a porosity of 79% and average pore sizes of 500 μm.

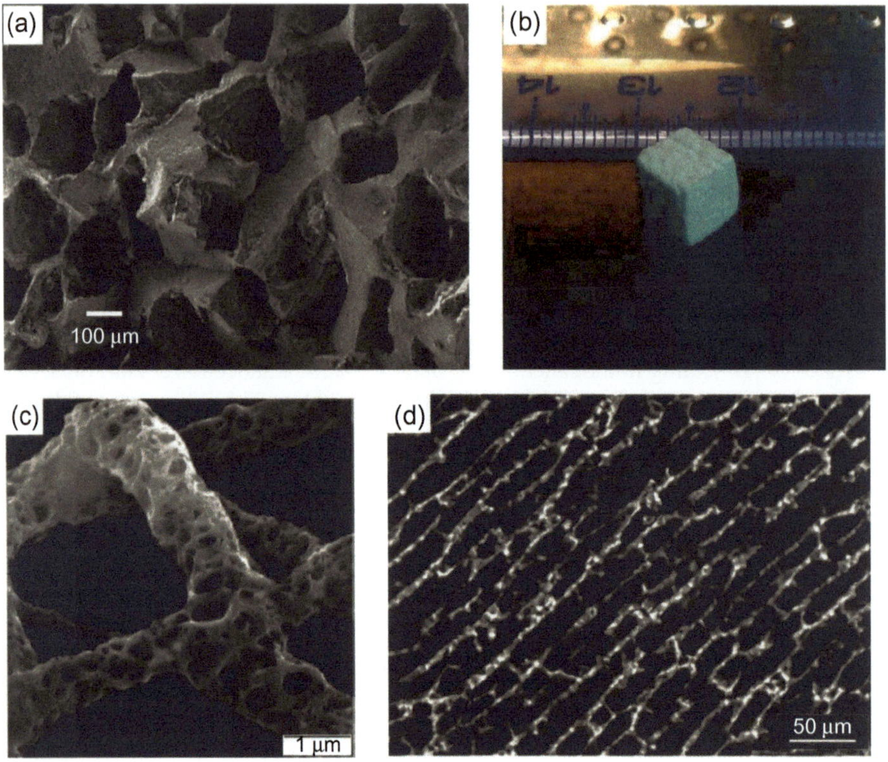

Figure 3.3 (a) Porous β-TCP scaffold fabricated by spark-plasma sintering of nanocrystalline powder and using 350–650 μm polyethylene glycol particles as porogens (adapted with permission from Ref. [38]); (b) a porous HA scaffold prepared by gas-foaming technique; (c) interconnected nanoporous network of calcium phosphate prepared by electrospinning of phosphate precursor sol mixed with PVC and subsequent calcination (adapted with permission from Ref. [39]); (d) porous HA scaffold fabricated by freeze-casting for bone tissue engineering applications (adapted with permission from Ref. [40]).

Electrospinning techniques. Electrospinning is a convenient fabrication strategy for fibrous 2-D or 3-D ceramic nanostructures with controllable microstructures [37,41]. The principle of electrospinning includes: (1) charge polymeric solutions containing metallic precursors or ceramic particles with a high voltage and (2) the charged solution is drawn by electric field from a nozzle onto a collector to form desirable structures (illustrated in the lower part of Figure 3.2). Many ceramic materials can be electrospun to mimic the nanoscale architectures of biological systems, such as fibrous proteins assembled in the native extracellular matrix or collagen fibrils in bone. For example, calcium phosphate precursor sol mixed with polyvinyl chloride (PVC) was electrospun and then sintered into highly interconnected nanoporous, ceramic networks (Figure 3.3c), which may be used for treating bone defects [39].

Freeze-drying techniques. Freeze-drying and freeze-casting are efficient techniques to produce 3-D porous, complex-shaped nanoceramic with controllable porosity and large pore sizes within the micrometer range [37]. In these techniques, ceramic slurry is first cast into a mold and then frozen, where frozen solvent serves as a binder to hold the ceramic green body together. Subsequently, the green body is freeze-dried to allow the solvent to sublimate under vacuum, during which the solvent acts as a porogen to form porous structure. At last, the porous green body is sintered to form porous ceramic scaffold. Water is a common solvent and the growth kinetics of ice crystals can be readily controlled during freezing to obtain desired porosity, pore geometry, and orientation [40]. Figure 3.3d shows the porous HA scaffold fabricated by freeze-casting for bone tissue engineering applications.

Anodization techniques. Anodization has also been employed to fabricate 3-D nanoporous or nanotubular scaffolds of anodic titanium oxide and aluminum oxide [42,43]. These techniques are essentially the same as the anodization methods introduced in Chapter 2 and, thus, are not expanded upon here.

Combining the techniques mentioned above offers more powerful tools to fabricate nanoscaffolds with hierarchical, interconnected pores or more complex porous structures. For example, gas-forming and freeze-drying were used together to prepare nano-calcium phosphate (including nano-HA and nano-biophasic calcium phosphate) scaffolds with both large interconnected pores (e.g., $>100\,\mu m$) and micropores (pore sizes from nanometers to submicrons) [37].

3.2 Nanoceramics for orthopedic applications

Ceramic materials including nanoceramics used in orthopedic applications are generally categorized into three groups: bioinert, bioactive, and bioresorbable. Bioinert ceramics like Al_2O_3 and TiO_2 do not interact with host tissues and serve as stable, chemically inactive, but biocompatible implant materials. In contrast, bioactive ceramics such as HA and bioactive glass interact with host tissues to form a biological bond at the interface between material and biological system, where mass and energy transfers constantly occur after the interfacial bond established. Bioresorbable ceramics such as TCP and modified bioglass are biodegradable and gradually replaced by the host tissues. The degradation products, however, need to be nontoxic, biocompatible, or even bioactive. Nanoceramics in these three categories have been broadly studied and developed for orthopedic applications in forms of particles, coatings, and scaffolds.

3.2.1 Nanoparticles

3.2.1.1 Iron oxides

In the past two decades, iron oxide or superparamagnetic iron oxide nanoparticles (SPION, e.g., γ-Fe_2O_3, Fe_3O_4 and associated compounds) have been actively studied for medical image, drug delivery, and hyperthermia treatment purposes. SPION has great magnetic properties and is generally biocompatible, only showing toxic effect at high dosages and over a long time period [44,45]. Due to the prevalent uses of magnetic resonance imaging (MRI) and X-ray in orthopedics, SPION and its derivatives have been successfully developed as contrast enhancing agents with high efficacy.

There are plenty of studies on SPION-based imaging contrast agents for visualization of musculoskeletal tissues or implanted orthopedic materials. For example, a variety of SPION derivatives have been demonstrated as excellent MRI contrast agents for bone marrow imaging due to their ability to decrease transverse relaxation time constants ($T2^*$) [46]. SPION is also used to increase the radiopacity of bone filling and augmentation materials like HA or self-setting cements, facilitating the monitoring of implantation and early diagnosis of complications [47]. In addition, a recent study demonstrates that bisphosphonate-conjugated SPION can target bone and track dynamic bone turnover, opening a new approach for diagnosis and monitoring of metabolic bone diseases and related bone pathology [48].

For many bone diseases such as osteoporosis, osteoarthritis, and bone cancer, SPION has been modified and studied for possible treatment of the bone diseases and/or simultaneously promotion of bone growth at the lesions [49]. Specifically, biocompatible SPION is conjugated with specific drug molecules, which can be delivered to disease sites with the assistance of directional magnetic fields. The drug molecules include those for treating bone diseases such as bis- and di-phosphonates and estrogen modulators, and bioactive peptides or bone growth factors for promoting bone regrowth. For example, Balasundaram and Webster [50] developed HA-coated SPION and functionalized the nanoparticles with bioactive YRGDSPC peptide by silane chemistry. The modified SPION retained its magnetic properties for directional delivery to osteoporosis sites under a magnetic field (Figure 3.4a). A follow-up study revealed that calcium phosphate-coated γ-Fe_2O_3 (Figure 3.4b) in the presence of bovine serum albumin significantly increased osteoblast proliferation compared to the controls without nanoparticles, revealing the effect of surface-modified SPION on promoting bone growth [51].

Other material properties of SPION also set up new and promising scenarios for treating bone-related cancer and infection without using anticancer drugs and antibi-

Figure 3.4 (a) Directional aggregation of HA-coated SPION functionalized with bioactive peptide under a magnetic field. (b) Transmission electron microscopy (TEM) image of calcium phosphate coated γ-Fe_2O_3 nanoparticles, scale bar = 200 nm.
Modified from Refs. [50,51] with permission.

otics. For cancer treatment, strategies such as thermal ablation and magnetic hyperthermia have been developed and preclinically tested [52,53]. These strategies utilize the heat generated by SPIONs in an oscillating magnetic field to destruct cancer cells *in situ*. Since SPIONs can be used as image contrast enhancing agents, a new strategy of integrating imaging and treatment functions together at the same lesion site is developed, in which the SPIONs act as a contrast enhancing agent first and subsequently kill the cancer cells by generating heat in the oscillating magnetic field. For treating infection that is commonly associated with implant surgery, surface-modified magnetic nanoparticles demonstrate strong inhibitory effects against bacterial and biofilm without adding antibiotics. For example, SPIONs incorporated with different metal ions such as iron, zinc, and silver have shown strong reduction effects against *Staphylococcus aureus* biofilm by penetrating to the bacterial cells (Figure 3.5) [54]. In contrast, conventional antibiotics including penicillin, oxacillin, gentamicin, and streptomycin are only effective against free-floating planktonic bacteria. Similarly, SPION in combination with fructose metabolites has been reported as a simple, broad-spectrum, and low-cost alternative to existing antibacterial strategies [55]. The fructose modified SPION reveals efficacy against both gram-positive and gram-negative bacteria infections and outperforms vancomycin, which is considered to be the antibiotic of last resort.

3.2.1.2 Calcium phosphates

Due to their prevalent applications and familiarity in the orthopedic field, calcium phosphates probably draw the most attention from the scientists and engineers in the arena of nanobiomaterials. Different sorts of calcium phosphate nanoparticles or nanosized powders are broadly used as base materials to prepare bioactive coating for orthopedic implants, paste or cement for filling bone voids, and freestanding scaffold for bone substitution. Besides, nanoparticles of HA, TCP, calcium tetraphosphate ($Ca_4P_2O_9$), and their derivatives have also been investigated for delivering drugs or molecules with delicate control of release kinetics. For instance, Cai et al. designed and fabricated hollow HA nanospheres, which could perform an accurate "on–off" release of drugs triggered by ultrasound [56]. The hollow HA nanospheres were synthesized by a wet-chemical route using hexadecyl (cetyl) trimethylammonium bromide as a modifier. These uniform and monodispersed hollow nanospheres can collapse and transfer into pin-shaped HA nanocrystallites triggered by an ultrasound wave and regulated by altering ultrasound power density. In addition, a demonstration study using amylose indicated that the drug release amount could be consistently controlled by ultrasound application time [56,57].

Calcium phosphate nanoparticles are also novel nonviral vectors for gene delivery [58]. Many studies have demonstrated that nanometer calcium phosphates possess higher penetration rates into cell membrane and their transfection efficiency can be up to 25-fold higher than that of conventional particles. Recently, dye-loaded calcium phosphate nanoparticles have been developed for photodynamic therapy against bacteria and cells. The strategy of photodynamic therapy utilizes biocompatible light-sensitive substances (i.e., photosensitizers) that become toxic when exposed to

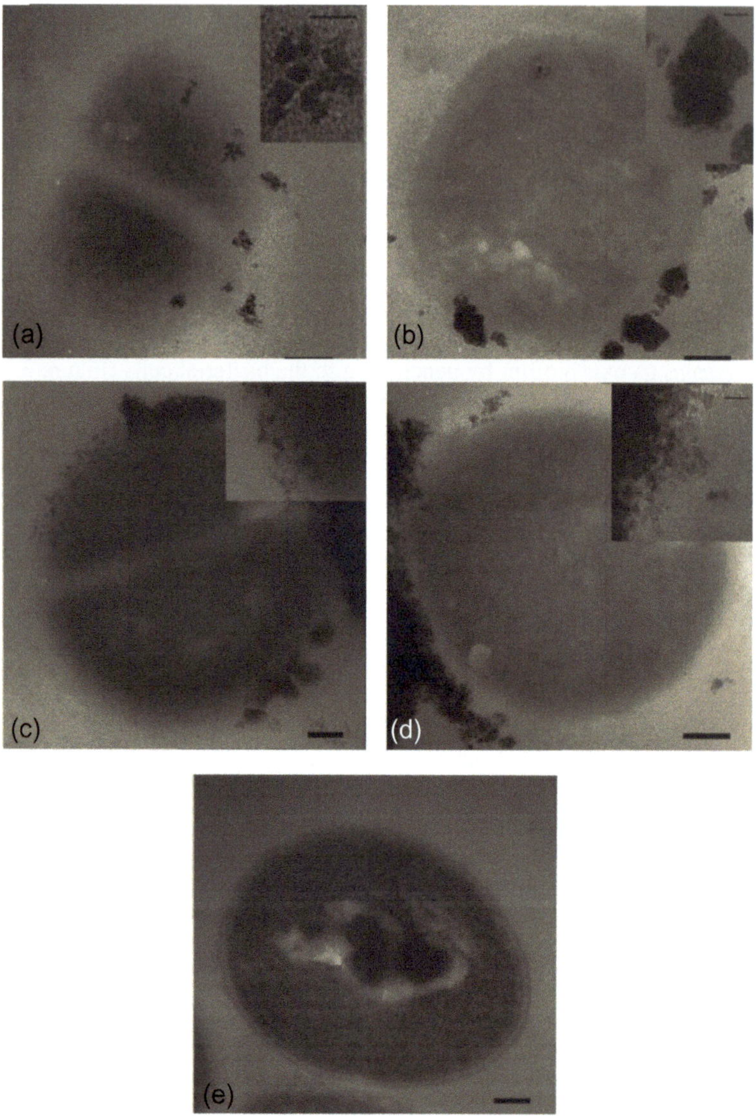

Figure 3.5 TEM images of SPION interacting with *S. aureus* membrane surface, which is related to growth inhibition and reduction of biofilm. Conjugated SPION including iron (a), silver (b), zinc (c), and chelated SPION (d) adsorb to the surface of *S. aureus*, while no SPION control is shown in (e), scale bars = 100 nm in all images. SPION also penetrate the membrane surface, as shown in insets (inset scale bars = 20 nm).
Reproduced with permission from Ref. [54].

specific lights to target and destroy malignant or diseased cells. Calcium phosphate nanoparticles have been studied as efficient carriers for the incorporation and delivery of photosensitizers (such as 5,10,15,20-tetrakis(3-hydroxyphenyl)-porphyrin) against cells (e.g., synoviocyte, macrophage) and bacteria (e.g., *S. aureus*) [59,60].

There is an increasing interest in studying the impact of calcium phosphate nanoparticles on directing the fate of bone mesenchymal stem cells (BMSCs). It has been revealed that crystallinity, microstructure and other chemical or physical properties of nanoparticles are pertinent to the stem cell fate [61,62]. Many studies have reported that HA with smaller sizes, preferably within the nanometer range, has a greater effect on enhancing BMSC adhesion and proliferation. In addition, adhesion, proliferation, and osteogenic differentiation of BMSCs are enhanced on well-crystallized (i.e., high crystallinity) nanophase HA compared to that of poor crystallinity. A mechanistic study by Liu et al. [63] suggests that concentrations of the degradation products of calcium phosphates (including extracellular calcium and inorganic phosphate) play an important role in mediating BMSC apoptosis, proliferation, and differentiation.

However, there is both research and clinical evidence showing nanometer particles released from calcium phosphates may largely impair the ability of BMSCs to proliferate and mature into a functional osteoblast phenotype [64]. This opposite evidence indicates that more and thorough research on utilizing stem cell and calcium phosphate nanoparticle for bone repair and regeneration is required before possible clinical applications.

3.2.1.3 Other ceramic nanoparticles

A few other ceramic nanoparticles have been studied to date for orthopedic applications, most of which, however, are used as additives to other orthopedic materials. For example, bare or functionalized magnesium oxide, zirconia, barium sulfate, and calcium carbonate are added to polymethylmethacrylate (PMMA) bone cement to reduce the exothermic effect of PMMA while increase its cytocompatibility, X-ray radiopacity, as well as antibacterial potential [65].

A comparison study between $BaSO_4$ microparticles and nanoparticles as additives in bone cement showed that substitution of nanoparticles for microparticles resulted in a 41% increase in tensile strain-to-failure, a 70% increase in tensile work-of-fracture, and a twofold increase in fatigue life of the bone cement [66]. These results indicate that nanoparticulate substitution of radiopacifiers substantially improved the ductile behavior and toughness of PMMA bone cement. In addition, $BaSO_4$ and ZrO_2 nanoparticles functionalized with 3-(trimethoxysilyl)propyl methacrylate revealed promising effects on reducing cytotoxicity and brittleness of PMMA bone cement compared to unfunctionalized nanoparticles or conventional $BaSO_4$ and ZrO_2 particles [67]. Recently, antibacterial effects of $BaSO_4$ nanoparticles against *Staphylococcus aureus* and *Pseudomonas aeruginosa* have been discovered, suggesting their potential applications as anti-infective additives to bone cement, implant coating, and medical tubing [68].

Calcium carbonate nanoparticles are also added to PMMA bone cement to increase its mechanical properties. A study by Hill et al. evaluated the efficacy of adding calcium carbonate nanoparticles on improving the mechanical properties of PMMA bone cement. Their study concluded that an optimized condition could provide improvements of 21% in stain energy at maximum load, 10% in Young's modulus, 7% in bending strength, and 8% in bending modulus of the bone cement when compared to the bone cement without adding calcium carbonate nanoparticles [69].

3.2.2 Nanocoatings

Ceramic nanocoatings on bone implants or scaffolds can be categorized as protective or functional coatings, serving for different purposes depending on the location in the musculoskeletal system they used. Protective nanocoating generally provide chemical or mechanical shields against corrosion, impact, abrasion, or material fatigue, while functional nanocoating mainly improve the biological performance or bioactivity (for orthopedics, mostly osseointegration and anti-infection properties) of underlying material at the material–bone interface. One of the research focuses is on the balance between protective and functional properties of orthopedic nanocoatings, aiming at the improvement of both lifetime and efficacy of orthopedic devices. Another important trend of nanocoating research is moving toward multipurpose, multifunctional, and smart surface with capabilities of tissue regeneration, biodetection and sensing, drug or molecule delivery, and so on. In this section, nanocoatings based on oxides, phosphates, and apatite are introduced and reviewed.

3.2.2.1 Oxides

This category of orthopedic nanocoatings typically includes Al_2O_3, TiO_2, and ZrO_2, which have been investigated for more than a decade. For example, Briggs et al. [70] devised nanoporous Al_2O_3 coating on Ti-based substrate for orthopedic use more than a decade ago. They deposited Al coating on metallic substrate by electron-beam evaporation and anodized the Al layer to obtain nanoporous Al_2O_3 coating at 160 V in phosphoric acid. Figure 3.6 shows the finishing of the Al_2O_3 coating on Ti6Al4V

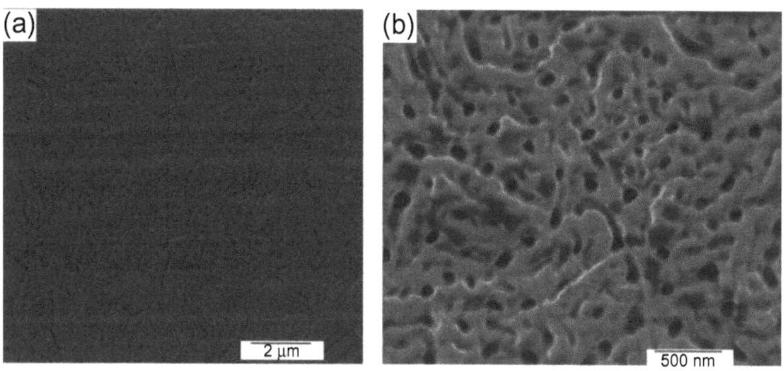

Figure 3.6 SEM images of Al_2O_3 coatings on Ti6Al4V substrate by anodization: (a) low-magnification image showing large grains in the coating and (b) high-magnification image showing nanometer pores in the coating as for (a).
Reproduced with permission from Ref. [70].

substrate, revealing nanoscale pores of 100–200 nm in diameter. This study also evaluated mechanical and cytocompatibility properties of the Al_2O_3 nanocoating. Results from shear and tension tests showed more than 20 MPa in shear strength and 10 MPa in tensile strength, suggesting a robust adhesion between the nanocoating and the substrate. Also, encouraging cell response on Al_2O_3 coatings revealed a significantly faster osteoblast growth on such coatings compared to cell culture coverslips (Thermanox). Furthermore, these researchers also incorporated other bioactive material such as colloidal silica to the Al_2O_3 nanocoatings to further improve its biological performance [71].

The effect and mechanism(s) of enhanced osteoblast growth on nanostructured or nanophase Al_2O_3 coatings have been verified and probed. By far, many studies indicate that appropriate surface roughness mimicking natural bone and select protein adsorption (e.g., vitronectin adsorption) are probably two key factors affecting cell or tissue responses [72]. A study by Webster et al. [73] found that enhanced adsorption of vitronectin on nanophased Al_2O_3 was possibly a result of decreased adsorption of apolipoprotein A-I and/or increased adsorption of calcium on nanophased Al_2O_3. For Al_2O_3 nanocoatings prepared by anodization, a recent investigation elucidated a correlation between topography-induced changes of surface electrochemical state and protein adsorption as well as osteoblast adhesion [74]. The findings of the study establish an approach to develop bioactive anodic Al_2O_3 nanocoatings by tailoring zeta potential and isoelectric point of the coating.

Although Al_2O_3 nanocoating exhibits high shear or tensile strength and improved bioactivity, the intrinsic brittleness of Al_2O_3 and its low resistance to crack propagation have restricted its use in clinical settings. Table 3.1 shows bulk mechanical properties of different ceramics including several nanoceramics. In the 1990s, yttria-stabilized ZrO_2 (Y-TZP) became a popular alternative to Al_2O_3 as a structural bioceramic due to its mechanical properties outperforming Al_2O_3 (Table 3.1). Recently, engineers have

Table 3.1 **Bulk mechanical properties of bioceramics and human cortical bone**

Ceramic material	Toughness (K_{IC}, MPa m$^{1/2}$)	Threshold stress intensity factor (K_{I0}, MPa m$^{1/2}$)	4-point bending strength (MPa)
Al_2O_3	4.2	2.4	400–600
ZrO_2	5.4	3.5	1000
HA	0.9	0.6	50–60
Nano-HA (grain size 50 nm)	1.5	–	–
TCP	1.3	0.8	50–60
Y_2O_3-stablized ZrO_2 (Y-TZP)	10.5	–	1000
Micro–nano-Al_2O_3–ZrO_2	6	5	600
Nano–nano-Ce-TZP–Al_2O_3	8.4	4.6	900
Human cortical bone	2.2–9.0	–	140–190

developed nanotechnology-enhanced ZrO_2 toughened Al_2O_3 (ZTA) to meet the clinical needs. Two examples shown in Table 3.1 are micron-grained Al_2O_3 embedded with ZrO_2 nanoparticles as a reinforced phase (i.e., micro–nano-Al_2O_3–ZrO_2 composite) and nanocomposite consisting of nanophase Al_2O_3 and Ce-stabilized ZrO_2 (nano–nano-Ce-TZP–Al_2O_3). These two types of ZTA have significantly higher toughness and strength values than Al_2O_3 or ZrO_2 alone. However, these nanocomposites have yet been used as coatings of orthopedic implants but as structural components like acetabular cup, femoral head, and liner, which will be introduced in Section 3.2.3.

Another important oxide with great interest in orthopedics is titania (TiO_2 the native oxide of Ti). As introduced in Chapter 2, Ti implant in the air has a thin oxide layer of a few nanometers thick on the surface and this TiO_2 layer directly interfaces with tissue or cells *in vivo*. In this sense, all Ti implants are practically coated with TiO_2 nanocoatings. The functions of nanotubular or nanoporous TiO_2 in promoting bone cell or tissue responses have been reviewed in the previous chapter. An exemplary work of dense TiO_2 ceramic surfaces with varied grain sizes is discussed here [75]. As shown in Figure 3.7, osteoblast adhesion on such TiO_2 nanocoating surfaces revealed significant changes, and the TiO_2 nanocoatings with grain sizes below 32 nm showed the highest adhesion efficiency. The osteoblast adhesion, however, is lower on TiO_2 nanocoatings with grain sizes of 97 and 56 nm compared to the coatings with grain sizes below 32 nm. These results indicate the surface topography (correlated to grain size) of TiO_2 nanocoating needs to be deliberately tailored for improving the biological responses to implants.

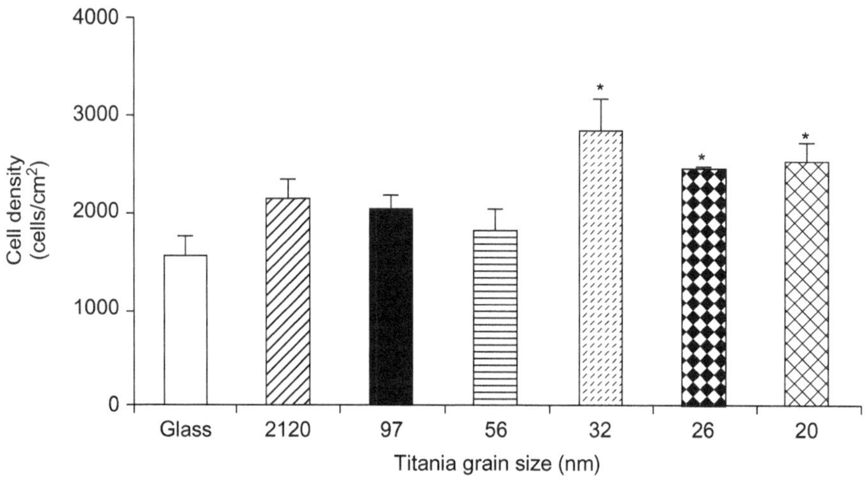

Figure 3.7 Osteoblast adhesion on titania of various grain sizes. Rat calvarial osteoblasts (3500 cell/cm²) in Dulbecco's modified Eagle medium (DMEM), containing 10% fetal bovine serum, were allowed to adhere on the surfaces of various grain size titania and borosilicate glass (reference material) under standard cell culture conditions (37 °C, humidified, 5% CO_2/95% air environment) for 4 h. Values are mean ± SEM; $n=3$; $*p<0.01$ (compared to reference glass).
Modified with permission from Ref. [75].

3.2.2.2 Calcium phosphates and apatite

As listed in Table 3.1, calcium phosphate ceramics (e.g., HA, TCPs, etc.) are mechanically weak and exhibit poor crack growth resistance compared to oxides like Al_2O_3 and ZrO_2. These intrinsic drawbacks limit their uses to only non-load-bearing applications, typically osteoconductive coatings on metallic prosthesis. For example, conventional HA coatings have been applied to commercial total hip implants and the HA-coated implant stem has demonstrated high efficacy on improving the bone–implant bonding and osseointegration for elderly patients or those who have less bone remodeling ability. Successful clinical use of conventional HA coating in orthopedic implants builds confidence for biomaterial engineers to further improve calcium phosphate mechanical performance and biological efficacy through nanotechnology.

In terms of fabrication, the calcium phosphate or apatite nanocoatings are deposited on the metallic implant surfaces or other substrates by a variety of electrochemical, wet-chemical, physical vapor deposition (PVD), or chemical vapor deposition (CVD) methods. For example, nanograined calcium phosphate coatings can be electrochemically deposited on Ti alloy substrates using aqueous electrolyte maintained at acidic pH, and the phase composition, grain size, and structure can be further modified by using different cathodic current densities and/or ultrasonic agitation [76]. The effective wet-chemical coating methods for fabricating calcium phosphate or apatite-based nanocoatings include the sol–gel process [25], coprecipitation [77], and *in situ* mineralization [78] routes. Other common techniques for preparing calcium phosphate nanocoatings are plasma spray deposition, ion beam-assisted deposition, pulsed laser deposition, and magnetron sputtering, whose details can be found in the previous chapters or referred to in the review article by Paital and Dahotre [79].

Calcium phosphate and apatite nanocoatings exhibit at least two major advantages over conventional, micron-sized coatings. First, mechanical strength and fracture toughness of the nanocoating can be much higher than those of the micron-sized coatings. This enhancement is essentially attributed to highly compact, nanosized grains in the coating, which usually results from the sintering of ultrafine calcium phosphate powders or nanocrystals pre-fabricated by the processes like wet-chemical routes. Mechanistically, the ultrafine calcium powders used for sintering nanostructured coatings can improve sinterability and enhanced densification due to their greater surface area-to-volume ratio (higher specific surface area), leading to dense, fine-grained microstructure in the nanocoatings [80]. For example, nanosized HA powder was sintered and hot-pressed to form a dense nanostructured ceramic with mean grain sizes below 100 nm, revealing increased mechanical properties [81]. When the grain size decreased from 3 μm to 50 nm, the fracture toughness and microhardness of HA ceramic increased from 0.28 to 1.52 MPa $m^{1/2}$ and 2.39 to 4.30 GPa, respectively. The increased toughness value is also greater than the highest toughness value of micron HA reported by others.

Second, calcium phosphate nanocoatings possess better biocompatibility and osteoconductive properties than micron-sized coatings. Almost all kinds of calcium phosphates studies by far have exhibited such improvement, including HA, TCP, dicalcium phosphate dehydrate (DCPD or brushite, $CaHPO_4{\cdot}2H_2O$), and so on [82,83]. The mechanism(s) behind the significantly enhanced biocompatibility and osteoconductivity for calcium phosphate nanocoatings has not been completely understood. However, a general consensus is that the enhancement is a result of many synergetic effects of nanotopography, hydrophobicity, surface chemistry, and surface charge, some of which have been discussed in the previous sections. For example, creating nanoscale rough surfaces has been shown to alter surface wettability and improve protein adsorption (such as vitronectin and fibronectin), leading to increases in subsequent osteoblast attachment and osseointegration compared to micron rough surfaces [84]. Although mechanistic studies are further needed, many researchers believe that the performance of calcium phosphate nanocoatings will achieve clinical success. For example, Chevalier and Gremillard predicted that the needs for tunable bioactive ceramics are soaring for the next few years and clinical uses of calcium phosphate nanocoatings are expected in the 2010s [85].

3.2.3 Nanotechnology-enhanced structural ceramics

Ceramics are well-known structural materials for building articulating surfaces in total joint implants (e.g., hip and knee) and the commercial products of Al_2O_3, ZrO_2, ZTA, and TZP are nowadays available globally. Since 1990, there have been more than 3.5 million Al_2O_3 components and more than 600,000 ZrO_2 femoral heads implanted worldwide [85]. However, these ceramic materials still suffer from potential problems of fracture, fatigue, aging, degradation, and wear during their service after implantation; moreover, other long-term risks associated with these ceramics are still unclear. For example, Al_2O_3 and ZrO_2 are intrinsically brittle (i.e., low in toughness), while TZP and ZTA experience the risk of low temperature degradation due to the meta-stability resulted from stress-induced phase transformation in ZrO_2. Therefore, ceramics with better performance are highly wanted. Nanoceramics and ceramic nanocomposites become promising candidates with an anticipation of entering clinical use within the next 10 years [85]. These nanomaterials possess enhanced properties that are attractive to orthopedic uses, including higher hardness, better wear behavior, and improved crack resistance and toughness compared to conventional, microstructured ceramics.

Among a large variety of possible systems, nanocomposites based on Al_2O_3 and/or ZrO_2 system are the most advanced ones in terms of manufacture and well-established biocompatibility. These nanocomposites are categorized to two groups: nanoparticle-embedded composite (referred to as "micro–nano-composites") and nanocomposite composed of nanoscale constituents (referred to as "nano–nano-composites"), as illustrated in Figure 3.8. Specifically, the micro–nano-composite is a composite in which ZrO_2 nanoparticles are evenly dispersed in micron Al_2O_3 grains, whereas the nano–nano-composite has both Al_2O_3 and ZrO_2 phases with grain sizes less than 500 nm. Both systems aim at increasing the resistance to crack propagation, tensile strength, and material stability of the ceramics, although the mechanisms behind the reinforcements are different. Specifically, micro–nano-composite benefits from the extremely large residual stress around ZrO_2 nanoparticles embedded in the Al_2O_3 grains, rendering the toughening of Al_2O_3 ceramic possible [86]. The strengthening mechanism of nano–nano-composite, in contrast, is reducing grain sizes and maintaining fully dense structure.

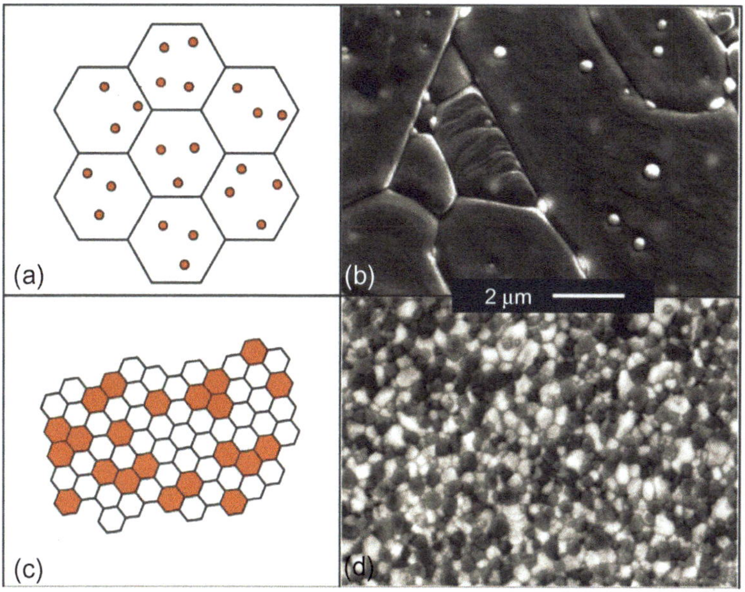

Figure 3.8 Schematics of (a) nanoparticle-embedded composite (micro–nano-composites) and (c) nanocomposite composed of nanoscale constituents (nano–nano-composite); SEM images of (b) micro–nano-Al_2O_3–ZrO_2 composite and (d) nano–nano-Al_2O_3–ZrO_2 composite. Reprinted with permission from Ref. [85].

Figure 3.9 plots the crack propagation resistance curves of several bioceramics including the ceramic nanocomposites mentioned above. It is very clear that both micro–nano-Al_2O_3–ZrO_2 composite and nano–nano-Ce-TZP/Al_2O_3 composite have superior crack propagation resistance (reflected by higher stress intensity factor K_I) over Al_2O_3, 3Y-TZP, and A10Z0Y (ZTA with 10 vol.% yttria-free zirconia) at any crack propagation speeds tested. Also, the nano–nano-Ce-TZP/Al_2O_3 composite appears to have even better crack propagation resistance than the micro–nano-Al_2O_3–ZrO_2 composite. In addition, Table 3.1 lists the toughness and bending strength of two nanocomposites. In comparison to other bioceramics, two nanocomposites exhibit higher and, more importantly, balanced toughness and strength. These comparisons clearly demonstrate the improvements in mechanical properties of structural bioceramics (especially toughness and strength) by employing nanotechnology to redesign the ceramics. However, it is also important to remind that wear and fatigue properties of the ceramic nanocomposites as implants have yet been reported and there are so far no long-term clinical studies on these materials. Obviously, these studies are necessary for future orthopedic applications of the nanocomposites.

3.2.4 Nanoscaffolds

In the present orthopedic clinical applications, 3-D scaffolds are mainly used as permanent structural supports to create a mechanical stable environment within bone defects

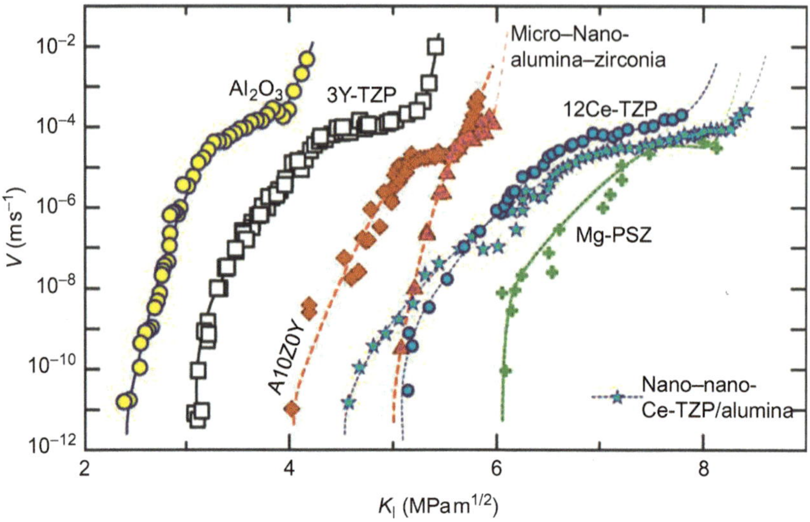

Figure 3.9 Crack propagation resistances of alumina, 3Y-TZP, 12Ce-TZP, Mg-PSZ (Mg-doped ZrO_2), A10Z0Y (ZTA with 10 vol.% yttria-free zirconia), micro–nano-Al_2O_3–ZrO_2 composite, and nano–nano-Ce-TZP/Al_2O_3 composite.
Reprinted with permission from Ref. [85].

or voids. For example, vertebral spacers and cages used in spinal fusion surgery are such scaffolds used to restore the height of fractured vertebra or maintain the satisfied space between neighboring vertebrae. Since the development of tissue engineering and regenerative medicine in the early 1990s, bioactive and bioresorbable scaffolds have been designed to control and direct cellular behavior and tissue growth by creating a biomimetic environment. A representative example is calcium phosphate scaffolds which mimic both natural bone structure and chemical composition. The calcium phosphate scaffolds can provide initial structural integrity for bone cells to attach and also facilitate their proliferation and differentiation as well as ultimate bone ingrowth. For orthopedic applications, 3-D scaffolds need to have open, interconnected pores with macropore sizes >100 μm, allowing bone cell ingrowth and remodeling [87]. In addition, injectable or gel-like scaffolds with proper porosity and osteophilic natures are also highly wanted in treating bone voids, defects, or fractures. When nanotechnology became available, an urgent need of developing nanoscaffolds that better mimic the hierarchical architecture of bone, has emerged. Undoubtedly, nanotechnology has been playing an important role in the design, fabrication, and modification of such scaffolds.

Ceramic nanoscaffolds possess many structural advantages that early generations of scaffolds don't have, making them robust systems for bone repair and regeneration. These advantages include high structural stability and mechanical properties, nanoscale porosity, high area-to-volume ratios, and high specific surface area. For example, macroporous β-TCP nanoscaffold with ultrafine grains (size about 200 nm) demonstrated improved compressive strength and elastic modulus of over 50–100% compared to the conventionally sintered scaffolds with micron-sized grains [38]. The nanoscaffold was fabricated from nanocrystalline powders by spark-plasma sintering (SPS) in a modified graphite die and its porosity (55–70%) and pore size (300–500 μm) could be readily controlled by altering the amount of porogen used (Figure 3.10). The compressive strength and elastic modulus of the nanoscaffolds with $66 \pm 1.0\%$ porosity were 3.1 ± 0.4 and 48.5 ± 8 MPa, respectively, while in contrast the conventional scaffold with the same porosity had much lower compressive strength (only 1.6 ± 0.5 MPa) and elastic modulus (31.2 ± 6 MPa). The enhanced mechanical properties were correlated to the microstructures of the nanoscaffold, which had ultrafine grains and increased density of solid matrix as a result of the enhanced neck growth of ceramic nanoparticles during the sintering process (Figure 3.10e vs. f).

Silicon carbide (SiC) is another nanomaterial that has been investigated for orthopedic prosthesis applications due to their high mechanical strength, good creep and oxidation resistances at elevated temperatures, chemical inertness, thermal stability, and resistance to corrosion [88]. 3-D porous SiC scaffolds have been fabricated and studied for potential applications in bone repair and regeneration. For example, in a recent study, osteogenic efficiency of bone cell ingrowth, proliferation, and early differentiation in a 3-D SiC porous ceramic scaffold with ultrasound treatment was studied [89]. Results showed that preosteoblast ingrowth depths inside the SiC scaffolds, average dsDNA content per scaffold, and alkaline phosphatase activity of cells were significantly increased in the SiC scaffold with ultrasound treatment compared to the sham control. This study reveals the great potential of SiC as a scaffold material for orthopedic clinical use and, therefore, SiC nanoscaffolds are expected to receive considerable attention in the near future. For example, SiC/HA nanocomposites have been developed recently for the fabrication of bone tissue engineering scaffolds. The nanocomposite of HA mixed with 5 wt% nano-SiC was sintered at 1100 °C or 1200 °C and it exhibited better mechanical and biological properties than pure HA [90].

In addition to structural advantages, nanotechnology provides ceramic scaffolds with new or advanced functions that are of great benefit for orthopedic diagnosis, imaging, surgery, and treatments such as drug or cell delivery. There is a trend in developing ceramic nanoscaffolds to serve multiple functions such as drug delivery, directing cell growth or tissue generation, and facilitating minimally invasive surgery. A recent study on scaffolds made by BONITmatrix®, a biphasic nanophase composite of silica and calcium phosphates (HA/TCP), showed a continuous release of gentamicin (an aminoglycoside antibiotic extensively used for orthopedics) from the scaffolds for 70 days without an initial burst release [91]. In addition, self-setting calcium phosphate cements (CPC) have been fabricated into injectable nanoscaffolds with high strength by incorporating chitosan and various porogens (such as absorbable fibers and mannitol), which strengthens the scaffolds and provides sequentially formed pores (for new bone growth), respectively [92]. The CPC nanoscaffolds demonstrated a controllable capability to deliver growth factors and proteins in a time-delayed manner by simply altering the CPC/chitosan ratio in the fabrication process. Besides, nanocrystalline

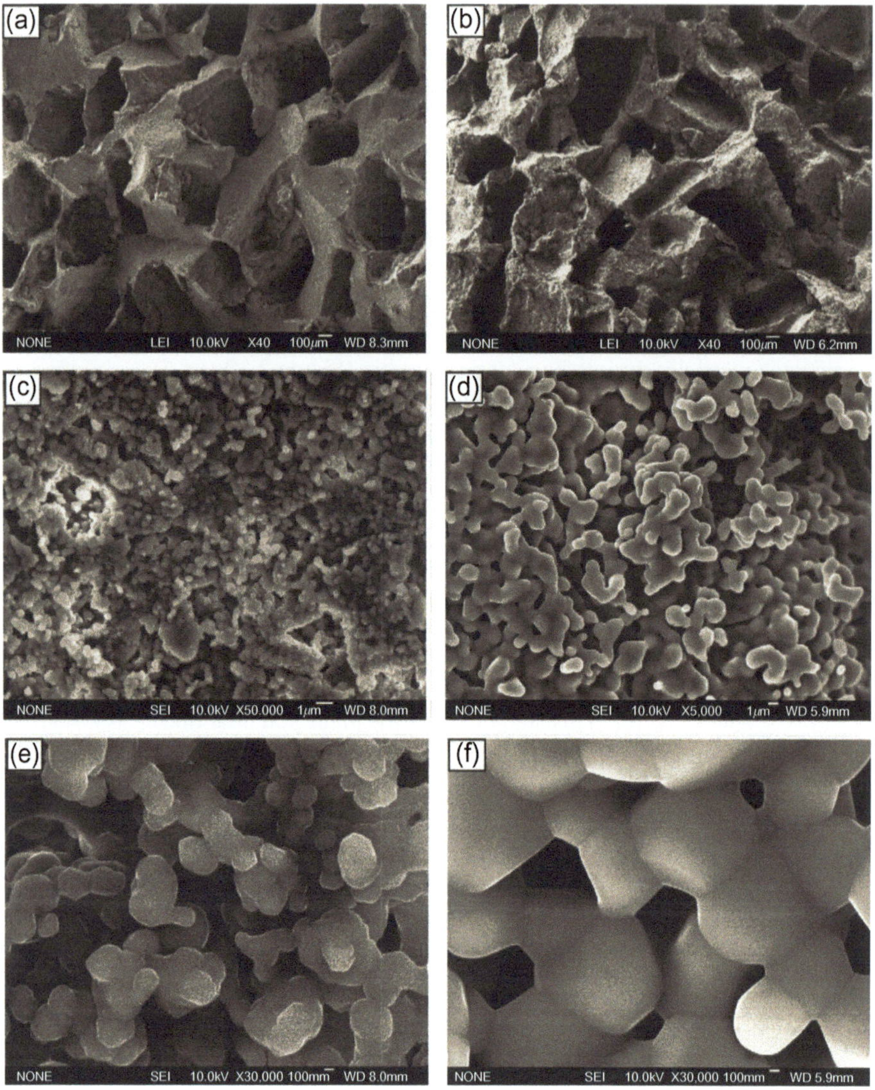

Figure 3.10 SEM images showing microstructures of (a, c, e) macroporous β-TCP nanoscaffold with ultrafine grain size ~200 nm prepared by SPS; and (b, d, f) β-TCP scaffold prepared by conventional sintering.
Reprinted with permission from Ref. [38].

HA/silica gel was studied as a tissue engineering scaffold for the delivery of BMSCs and platelet-rich growth factor and results indicated that this nanoscaffold-assisted strategy was more effective for bone regeneration *in vivo* compared to the treatment without nanoscaffold [93]. There are a lot of studies on composite nanoscaffolds composed of ceramics and other materials (mostly polymers or biomolecules) and Chapter 4 will deliver a comprehensive review on these nanocomposites.

3.3 Commercialization status of orthopedic nanoceramics

Compared to other nanomaterials, there are more nanoceramic products in the market, although majority of them are limited to ceramic nanoparticles. For research and development purposes, a variety of ceramic nanoparticles are already commercially available from chemical product companies like Sigma-Aldrich, but few of them are permitted or suitable for orthopedic applications. In the field of orthopedics, nanoceramic products are mainly developed and commercialized by small high-tech companies or start-ups. Some of the companies have already launched products to market, although the number of products is limited. In fact, most of the companies have products in different phases of clinical or preclinical trials and an increasing number of nanoceramic products are anticipated over the next few decades. In 2009, Chevalier and Gremillard already predicted that nanostructured oxide ceramics and calcium phosphate ceramics would be approved for clinical use within next 10 and 5 years, respectively [85].

For nanoceramics like calcium phosphates, the transfer of research outcomes from laboratory to clinical beds is relatively faster compared to other nanomaterials. To date, plenty of conventional ceramics have been successfully used in orthopedics for a long period of time with excellent or satisfactory safety and efficacy records. Therefore, it is expected that the nanostructured version of such ceramics may first become available for clinical use in the near future. Indeed, there are already some commercialized orthopedic nanoceramics that have reached or are close to product launch, which is summarized below.

A few companies have launched calcium phosphate-based nanoceramic products for spinal fusion, treatment of non-load-bearing bone defects, and anchoring artificial joints. For example, Etex Corporation, a U.S.-based company, has been developing and marketing its nanocrystalline calcium phosphate formulation as bone substitute materials for more than 10 years. So far its proprietary nanocrystalline technology has yielded a product line of at least three clinically proven products, including injectable cement and scaffold [94]. According to the information disclosed by Etex, at least two nanotechnology-based products have received U.S. Food and Drug Administration (FDA) clearance for treating bone voids that resulted from fracture and other medical conditions. Similar products based on nanocrystalline calcium phosphates have also been developed and approved for clinical use in China. A product line of bioactive bone substitute materials based on nanocrystalline calcium phosphate (combined with collagen in some products) has been developed by Prof. Fuzhai Cui at Tsinghua University, Beijing, and its core products have been approved by the Chinese FDA for clinical use as early as 2004 [95]. In India, according to a report from the Indian Advanced Center for Powder Metallurgy and New Materials, nanobioceramics for orthopedic and bone graft applications have been commercialized by a Chennai-based company, Eucare Pharmaceuticals [96]. In addition, a 3-year clinical study of the nanoscaffold composed of nano-HA/silica gel and developed by Artoss in Germany has been positively reported [97]. The same product has been recently proven to be effective in stem cell tissue engineering for bone regeneration by preclinical animal studies [93].

In contrast to calcium phosphate-based nanoceramics, the development of ZrO_2 transformation toughened nanoceramics such as nano-ZTA or -TZP has not yielded commercial orthopedic products in the market to date. However, as mentioned above, researchers are expecting the debut of commercial nano-ZTA or -TZP implants in the near future. Actually most of the raw materials and fabrication techniques for manufacturing these implants are already commercially available. For example, a company (INNOVNANO) has launched new grades of nano-zirconia Y-TZP (ZrO_2-3 mol.% Y_2O_3) and nano-ZTA that can be used for manufacturing nanocoatings by spraying techniques or nanograined implants by sintering [98]. Meanwhile, companies focusing on developing processing technology have commercialized a number of nanocoating or sintering techniques for scaling up the manufacture of such nanoceramics. A company (*n*Gimat) has successfully developed NanoSpray^{SM} Combustion Processing Technology for fabricating high performance coatings on various substrates [99].

Due to the maturation of fabrication materials and methods, ZrO_2 transformation toughened nanoceramics have immense potential of commercialization for orthopedic use. In fact, the European Nanoker integrated project, operated since 2005, is dedicated to commercializing structural ceramic nanocomposites of ZrO_2 and Al_2O_3 for resurfacing ceramic orthopedic implants such as hip resurfacing and unicondylar knee prostheses [85,100]. Prototypes of such resurfacing implants based on ceramic nanocomposites have been produced in the framework of the Nanoker project and launch of the products is planned.

3.4 Summary

Nanotechnology-enabled ceramic materials for orthopedic applications, including their fabrication techniques and commercialization status, have been analyzed in this chapter. To date, ceramic nanomaterials have exhibited immense potential in many aspects of orthopedic treatments (e.g., bone graft, joint replacement, bone filler, spinal spacing, etc.) and various forms of ceramic from nanoparticle to nanoscaffold have entered different stages of commercialization. In the near future, ceramic nanomaterials are expected to deliver more opportunities and advances in the areas of bone tissue engineering, minimally invasive surgery, and the next generation of implants and prostheses. Among these areas, improved implants and prostheses with better efficacy and lifetime are worth an emphasis because conventional orthopedic implants, such as Charnley based total hip implants will continue being widely used for next few decades. Also, these improved prosthetic implants will probably become a solution to the scenarios in which tissue engineering and minimally invasive surgery cannot solve the problems. To this end, the nanoceramics that would benefit the conventional implants in terms of enhancing performance, increasing safety, and prolonging functional lifetime are among the priorities of research and development. This trend is also reflected by the ongoing commercialization of structural ceramic nanocomposites which would advance the currently used orthopedic implant to a new stage.

References

[1] Brinker CJ, Scherer GW. Sol–gel science: the physics and chemistry of sol–gel processing. Access Online via Elsevier, 1990.

[2] Caruso RA, Antonietti M. Sol–gel nanocoating: an approach to the preparation of structured materials. Chem Mater 2001;13:3272–82.

[3] Gu F, Wang SF, Lü MK, Zhou GJ, Xu D, Yuan DR. Photoluminescence properties of SnO_2 nanoparticles synthesized by sol–gel method. J Phys Chem B 2004;108:8119–23.

[4] Xu J, Yang H, Fu W, Du K, Sui Y, Chen J, et al. Preparation and magnetic properties of magnetite nanoparticles by sol–gel method. J Magn Magn Mater 2007;309:307–11.

[5] Kuriakose TA, Kalkura SN, Palanichamy M, Arivuoli D, Dierks K, Bocelli G, et al. Synthesis of stoichiometric nano crystalline hydroxyapatite by ethanol-based sol–gel technique at low temperature. J Cryst Growth 2004;263:517–23.

[6] Hsu K-F, Tsay S-Y, Hwang B-J. Synthesis and characterization of nano-sized $LiFePO_4$ cathode materials prepared by a citric acid-based sol–gel route. J Mater Chem 2004;14:2690–5.

[7] Trewyn BG, Slowing II, Giri S, Chen H-T, Lin VS-Y. Synthesis and functionalization of a mesoporous silica nanoparticle based on the sol–gel process and applications in controlled release. Acc Chem Res 2007;40:846–53.

[8] Lu Y, Yin Y, Mayers BT, Xia Y. Modifying the surface properties of superparamagnetic iron oxide nanoparticles through a sol–gel approach. Nano Lett 2002;2:183–6.

[9] Deng Y-H, Wang C-C, Hu J-H, Yang W-L, Fu S-K. Investigation of formation of silica-coated magnetite nanoparticles via sol–gel approach. Colloids Surf A Physicochem Eng Asp 2005;262:87–93.

[10] Vallet-Regi M, Gonzalez-Calbet JM. Calcium phosphates as substitution of bone tissues. Prog Solid State Chem 2004;32:1–31.

[11] Elliott JC. Structure and chemistry of the apatites and other calcium orthophosphates. Amsterdam: Elsevier; 1994.

[12] Chai CS, Gross KA, Ben-Nissan B. Critical ageing of hydroxyapatite sol–gel solutions. Biomaterials 1998;19:2291–6.

[13] Gross KA, Chai CS, Kannangara GSK, Ben-Nissan B, Hanley L. Thin hydroxyapatite coatings via sol–gel synthesis. J Mater Sci-Mater Med 1998;9:839–43.

[14] Liu DM, Troczynski T, Tseng WJ. Water-based sol–gel synthesis of hydroxyapatite: process development. Biomaterials 2001;22:1721–30.

[15] Kim IS, Kumta PN. Sol–gel synthesis and characterization of nanostructured hydroxyapatite powder. Mater Sci Eng B-Solid 2004;111:232–6.

[16] Jillavenkatesa A, Condrate RA. Sol–gel processing of hydroxyapatite. J Mater Sci 1998;33:4111–9.

[17] Yang L, Ning XS, Jia WW, Chen KX, Zhou HP. Particle size control in sol–gel synthesis of nano hydroxyapatite. Adv Mater Res 2006;11–12:231–4.

[18] Bose S, Dasgupta S, Tarafder S, Bandyopadhyay A. Microwave-processed nanocrystalline hydroxyapatite: simultaneous enhancement of mechanical and biological properties. Acta Biomater 2010;6:3782–90.

[19] Yang L, Ning X, Bai Y, Jia W. A scalable synthesis of non-agglomerated and low-aspect ratio hydroxyapatite nanocrystals using gelatinized starch matrix. Mater Lett 2013;113:142–5.

[20] Iida H, Takayanagi K, Nakanishi T, Osaka T. Synthesis of Fe_3O_4 nanoparticles with various sizes and magnetic properties by controlled hydrolysis. J Colloid Interf Sci 2007;314:274–80.

[21] Lenggoro IW, Okuyama K, Fernandez de la Mora J, Tohge N. Preparation of ZnS nanoparticles by electrospray pyrolysis. J Aerosol Sci 2000;31:121–36.

[22] Wang Z, Zhang H, Zhang L, Yuan J, Yan S, Wang C. Low-temperature synthesis of ZnO nanoparticles by solid-state pyrolytic reaction. Nanotechnology 2003;14:11.

[23] Wang Y, Zhang S, Wei K, Zhao N, Chen J, Wang X. Hydrothermal synthesis of hydroxyapatite nanopowders using cationic surfactant as a template. Mater Lett 2006;60:1484–7.

[24] Tallon C, Moreno R, Nieto MI. Synthesis of γ-Al_2O_3 nanopowders by freeze-drying. Mater Res Bull 2006;41:1520–9.

[25] Ben-Nissan B, Milev A, Vago R. Morphology of sol–gel derived nano-coated coralline hydroxyapatite. Biomaterials 2004;25:4971–5.

[26] Roest R, Latella B, Heness G, Ben-Nissan B. Adhesion of sol–gel derived hydroxyapatite nanocoatings on anodised pure titanium and titanium (Ti_6Al_4V) alloy substrates. Surf Coat Technol 2011;205:3520–9.

[27] Guo L, Li H. Fabrication and characterization of thin nano-hydroxyapatite coatings on titanium. Surf Coat Technol 2004;185:268–74.

[28] Chai C, Ben-Nissan B. Bioactive nanocrystalline sol–gel hydroxyapatite coatings. J Mater Sci Mater Med 1999;10:465–9.

[29] Karthikeyan J, Berndt C, Tikkanen J, Reddy S, Herman H. Plasma spray synthesis of nanomaterial powders and deposits. Mater Sci Eng A 1997;238:275–86.

[30] Karthikeyan J, Berndt CC, Reddy S, Wang JÄ, King AH, Herman H. Nanomaterial deposits formed by DC plasma spraying of liquid feedstocks. J Am Ceram Soc 1998;81:121–8.

[31] Surmenev RA. A review of plasma-assisted methods for calcium phosphate-based coatings fabrication. Surf Coat Technol 2012;206:2035–56.

[32] Roy M, Bandyopadhyay A, Bose S. Induction plasma sprayed nano hydroxyapatite coatings on titanium for orthopaedic and dental implants. Surf Coat Technol 2011;205:2785–92.

[33] Rojaee R, Fathi M, Raeissi K. Electrophoretic deposition of nanostructured hydroxyapatite coating on AZ91 magnesium alloy implants with different surface treatments. Appl Surf Sci 2013;285:664–73.

[34] Farrokhi-Rad M, Ghorbani M. Electrophoretic deposition of titania nanoparticles in different alcohols: kinetics of deposition. J Am Ceram Soc 2011;94:2354–61.

[35] Mehdipour M, Afshar A, Mohebali M. Electrophoretic deposition of bioactive glass coating on 316L stainless steel and electrochemical behavior study. Appl Surf Sci 2012;258:9832–9.

[36] Tran RT, Naseri E, Kolasnikov A, Bai X, Yang J. A new generation of sodium chloride porogen for tissue engineering. Biotechnol Appl Biochem 2011;58:335–44.

[37] Hong Y, Fan H, Li B, Guo B, Liu M, Zhang X. Fabrication, biological effects, and medical applications of calcium phosphate nanoceramics. Mater Sci Eng: R 2010;70:225–42.

[38] Zhang F, Lin K, Chang J, Lu J, Ning C. Spark plasma sintering of macroporous calcium phosphate scaffolds from nanocrystalline powders. J Eur Ceram Soc 2008;28:539–45.

[39] Dai X, Shivkumar S. Hybrid analogs for the production of porous calcium phosphate scaffolds. Mater Sci Eng C 2008;28:336–40.

[40] Deville S, Saiz E, Tomsia AP. Freeze casting of hydroxyapatite scaffolds for bone tissue engineering. Biomaterials 2006;27:5480–9.

[41] Yang L, Sheldon BW, Webster TJ. Nanophase ceramics for improved drug delivery: current opportunities and challenges. Am Ceram Soc Bull 2010;89:24–31.

[42] Poinern GEJ, Shackleton R, Mamum S, Fawcett D. Significance of novel bioinorganic anodic aluminum oxide nanoscaffolds for promoting cellular response. Nanotechnol Sci Appl 2011;4:11–24.

[43] Raja K, Misra M, Paramguru K. Formation of self-ordered nano-tubular structure of anodic oxide layer on titanium. Electrochim Acta 2005;51:154–65.
[44] Pisanic II TR, Blackwell JD, Shubayev VI, Finones RR, Jin S. Nanotoxicity of iron oxide nanoparticle internalization in growing neurons. Biomaterials 2007;28:2572–81.
[45] Lewinski N, Colvin V, Drezek R. Cytotoxicity of nanoparticles. Small 2008;4:26–49.
[46] Na HB, Song IC, Hyeon T. Inorganic nanoparticles for MRI contrast agents. Adv Mater 2009;21:2133–48.
[47] Jakhmola A, Anton N, Vandamme TF. Inorganic nanoparticles based contrast agents for X-ray computed tomography. Adv Healthc are Mater 2012;1:413–31.
[48] Panahifar A, Mahmoudi M, Doschak MR. Synthesis and in vitro evaluation of bone-seeking superparamagnetic iron oxide nanoparticles as contrast agents for imaging bone metabolic activity. ACS Appl Mater Interfaces 2013;5:5219–26.
[49] Yang L, Webster TJ. Nanotechnology controlled drug delivery for treating bone diseases. Expert Opin Drug Del 2009;6:851–64.
[50] Balasundaram G, Webster TJ. Applications of magnetic nanoparticles for the treatment of osteoporosis. Mater Res Soc Symp Proc 2007;1019, 1019-FF02-06.
[51] Pareta RA, Taylor E, Webster TJ. Increased osteoblast density in the presence of novel calcium phosphate coated magnetic nanoparticles. Nanotechnology 2008;19:265101.
[52] Mazzaglia PJ, Berber E, Siperstein AE. Radiofrequency thermal ablation of metastatic neuroendocrine tumors in the liver. Curr Treat Options in Oncol 2007;8:322–30.
[53] Arruebo M, Fernández-Pacheco R, Ibarra MR, Santamaría J. Magnetic nanoparticles for drug delivery. Nano Today 2007;2:22–32.
[54] Taylor EN, Kummer KM, Durmus NG, Leuba K, Tarquinio KM, Webster TJ. Superparamagnetic iron oxide nanoparticles (SPION) for the treatment of antibiotic-resistant biofilms. Small 2012;8:3016–27.
[55] Durmus NG, Taylor EN, Kummer KM, Webster TJ. Enhanced efficacy of superparamagnetic iron oxide nanoparticles against antibiotic-resistant biofilms in the presence of metabolites. Adv Mater 2013;25:5706–13.
[56] Cai Y, Pan H, Xu X, Hu Q, Li L, Tang R. Ultrasonic controlled morphology transformation of hollow calcium phosphate nanospheres: a smart and biocompatible drug release system. Chem Mater 2007;19:3081–3.
[57] Cai Y, Tang R. Calcium phosphate nanoparticles in biomineralization and biomaterials. J Mater Chem 2008;18:3775–87.
[58] Maitra A. Calcium phosphate nanoparticles: second-generation nonviral vectors in gene therapy. Expert Rev Mol Diagn 2005;5:893–905.
[59] Schwiertz J, Wiehe A, Gräfe S, Gitter B, Epple M. Calcium phosphate nanoparticles as efficient carriers for photodynamic therapy against cells and bacteria. Biomaterials 2009;30:3324–31.
[60] Klesing J, Wiehe A, Gitter B, Grafe S, Epple M. Positively charged calcium phosphate/polymer nanoparticles for photodynamic therapy. J Mater Sci: Mater Med 2010;21:887–92.
[61] Hu Q, Tan Z, Liu Y, Tao J, Cai Y, Zhang M, et al. Effect of crystallinity of calcium phosphate nanoparticles on adhesion, proliferation, and differentiation of bone marrow mesenchymal stem cells. J Mater Chem 2007;17:4690–8.
[62] Kaivosoja E, Barreto G, Levón K, Virtanen S, Ainola M, Konttinen YT. Chemical and physical properties of regenerative medicine materials controlling stem cell fate. Ann Med 2012;44:635–50.
[63] Liu YK, Lu QZ, Pei R, Ji HJ, Zhou GS, Zhao XL, et al. The effect of extracellular calcium and inorganic phosphate on the growth and osteogenic differentiation of mesenchymal stem cells in vitro: implication for bone tissue engineering. Biomed Mater 2009;4:025004.

[64] Saldana L, Sanchez-Salcedo S, Izquierdo-Barba I, Bensiamar F, Munuera L, Vallet-Regi M, et al. Calcium phosphate-based particles influence osteogenic maturation of human mesenchymal stem cells. Acta Biomater 2009;5:1294–305.

[65] Arora M, Chan EK, Gupta S, Diwan AD. Polymethylmethacrylate bone cements and additives: a review of the literature. World J Orthop 2013;4:67.

[66] Gomoll AH, Fitz W, Scott RD, Thornhill TS, Bellare A. Nanoparticulate fillers improve the mechanical strength of bone cement. Acta Orthop 2008;79:421–7.

[67] Gillani R, Ercan B, Qiao A, Webster TJ. Nanofunctionalized zirconia and barium sulfate particles as bone cement additives. Int J Nanomed 2010;5:1–11.

[68] Aninwene II GE, Stout D, Yang Z, Webster TJ. Nano-$BaSO_4$: a novel antimicrobial additive to pellethane. Int J Nanomed 2013;8:1197.

[69] Hill J, Orr J, Dunne N. In vitro study investigating the mechanical properties of acrylic bone cement containing calcium carbonate nanoparticles. J Mater Sci Mater Med 2008;19:3327–33.

[70] Briggs EP, Walpole AR, Wilshaw PR, Karlsson M, Pålsgård E. Formation of highly adherent nano-porous alumina on Ti-based substrates: a novel bone implant coating. J Mater Sci: Mater Med 2004;15:1021–9.

[71] Walpole A, Briggs E, Karlsson M, Pålsgård E, Wilshaw P. Nano-porous alumina coatings for improved bone implant interfaces. Mater Werkst 2003;34:1064–8.

[72] Liu H, Webster TJ. Nanomedicine for implants: a review of studies and necessary experimental tools. Biomaterials 2006;28:354–69.

[73] Webster TJ, Schadler LS, Siegel RW, Bizios R. Mechanisms of enhanced osteoblast adhesion on nanophase alumina involve vitronectin. Tissue Eng 2001;7:291–301.

[74] Pedimonte BJ, Moest T, Luxbacher T, von Wilmowsky C, Fey T, Schlegel KA, et al. Morphological zeta-potential variation of nanoporous anodic alumina layers and cell adherence. Acta Biomater 2014;10:968–74.

[75] Webster TJ, Siegel RW, Bizios R. Osteoblast adhesion on nanophase ceramics. Biomaterials 1999;20:1221–7.

[76] Narayanan R, Seshadri S, Kwon T, Kim K. Electrochemical nano-grained calcium phosphate coatings on Ti–6Al–4V for biomaterial applications. Scr Mater 2007;56:229–32.

[77] Li P. Biomimetic nano-apatite coating capable of promoting bone ingrowth. J Biomed Mater Res A 2003;66:79–85.

[78] Tas AC, Bhaduri SB. Rapid coating of Ti6Al4V at room temperature with a calcium phosphate solution similar to 10× simulated body fluid. J Mater Res 2004;19:2742–9.

[79] Paital SR, Dahotre NB. Calcium phosphate coatings for bio-implant applications: materials, performance factors, and methodologies. Mater Sci Eng: R 2009;66:1–70.

[80] Yeong K, Wang J, Ng S. Fabricating densified hydroxyapatite ceramics from a precipitated precursor. Mater Lett 1999;38:208–13.

[81] Veljovic D, Jokic B, Petrovic R, Palcevskis E, Dindune A, Mihailescu IN, et al. Processing of dense nanostructured HAP ceramics by sintering and hot pressing. Ceram Int 2009;35:1407–13.

[82] Kumta PN, Sfeir C, Lee D-H, Olton D, Choi D. Nanostructured calcium phosphates for biomedical applications: novel synthesis and characterization. Acta Biomater 2005;1:65–83.

[83] Webster TJ. Nanophase ceramics: the future orthopedic and dental implant material. In: Advances in chemical engineering, vol. 27. New York: Academic Press; 2001. p. 125–66.

[84] Yang L, Zhang LJ, Webster TJ. Nanobiomaterials: state of the art and future trends. Adv Eng Mater 2011;13:B197–217.

[85] Chevalier J, Gremillard L. Ceramics for medical applications: a picture for the next 20 years. J Eur Ceram Soc 2009;29:1245–55.
[86] Chevalier J, Deville S, Fantozzi G, Bartolomé J, Pecharroman C, Moya J, et al. Nanostructured ceramic oxides with a slow crack growth resistance close to covalent materials. Nano Lett 2005;5:1297–301.
[87] Li X, Wang L, Fan Y, Feng Q, Cui FÄ, Watari F. Nanostructured scaffolds for bone tissue engineering. J Biomed Mater Res A 2013;101:2424–35.
[88] Mahmoodi M, Ghazanfari L. Fundamentals of biomedical applications of biomorphic SiC. In: Gerhardt R, editor. Properties and applications of silicon carbide. InTech: Open Access Publisher; 2011. p. 297–343.
[89] Wu L, Lin L, Qin Y-X. Enhancement of cell ingrowth, proliferation and early differentiation in a 3-D SiC scaffold using low-intensity pulsed ultrasound. Tissue Eng Part A 2014; 21(1–2): 53–61.
[90] Hesaraki S, Ebadzadeh T, Ahmadzadeh-Asl S. Nanosilicon carbide/hydroxyapatite nanocomposites: structural, mechanical and in vitro cellular properties. J Mater Sci Mater Med 2010;21:2141–9.
[91] Teller M, Gopp U, Neumann HÄ, Kühn KÄ. Release of gentamicin from bone regenerative materials: an in vitro study. J Biomed Mater Res Part B: Appl Biomater 2007;81:23–9.
[92] Xu HH, Weir MD, Simon CG. Injectable and strong nano-apatite scaffolds for cell/growth factor delivery and bone regeneration. Dent Mater 2008;24:1212–22.
[93] Behnia H, Khojasteh A, Kiani MT, Khoshzaban A, Mashhadi Abbas F, Bashtar M, et al. Bone regeneration with a combination of nanocrystalline hydroxyapatite silica gel, platelet-rich growth factor, and mesenchymal stem cells: a histologic study in rabbit calvaria. Oral Surg Oral Med Oral Pathol Oral Radiol 2013;115:e7–15.
[94] Etex product overview, https://secure.etexcorp.com/products_overview.html; 2014 [accessed 21.05.14].
[95] Bone substitute materials patented in the US. http://www.zgnmjs.com/news/10703736.html; 2011 [accessed 29.06.14].
[96] Purushotham H. Transfer of nanotechnologies from R&D institutions to SMEs in India. Carbon 2012;16:47.
[97] Heinemann F, Mundt T, Biffar R, Gedrange T, Goetz W. A 3-year clinical and radiographic study of implants placed simultaneously with maxillary sinus floor augmentations using a new nanocrystalline hydroxyapatite. J Physiol Pharmacol 2009;60:91–7.
[98] NNOVNANO. INNOVNANO launches new grades nanoparticles for zirconia coatings. Nanowerk; 2010.
[99] Jiang Y, White M, Venugopal G, Hunt AT. Development and commercialization of nanomaterials by the NanoSpraySM combustion processing technology. In: MRS proceedings. Cambridge University Press; 2013, p. mrsf12-1533-aaa04-02.
[100] World i. Structural ceramic nanocomposites for top-end functional applications, http://www.ist-world.org/projectdetails.aspx?projectid=fa0bae6730df4b4db3a9da774d292e15project description in IST world; accessed [26.07.13].

Bioinspired nanopolymers and nanocomposites for orthopedic applications

4

Biomedical macromolecules and polymers encompass both natural and synthetic substances that ranged from DNA, silk, and chitosan to synthetic polyesters. Well-known synthetic polymers such as poly(L-lactic acid) (PLLA), poly(lactic-co-glycolic) acid (PLGA), poly(dimethylsiloxane) (PDMS), poly(ε-caprolactone) (PCL), and polyurethane (PU) have already been widely used for clinical or laboratory researches. To date, many natural polymers (e.g., silk and collagen) and synthetic polymers (such as PLA and PLGA) have been approved by drug administrations in many countries for specific human clinical applications from surgical sutures to drug delivery systems. When nanotechnology emerged, polymers and polymer-based composites were among the earliest nanomaterials that were studied to enhance cell or tissue responses [1,2]. Owing to the prevalence of polymers and their composites in biomedical applications, their abilities to regulate cell or tissue activities have drawn great attention.

It is important to emphasize that many natural tissues are essentially composed of nanoscale biopolymers or biocomposites with hierarchical architectures. Therefore, by mimicking the structure and property of their natural counterparts, synthetic nanopolymers and nanocomposites are very likely to enhance/regulate the functions of specific cells or tissues. This principle has been demonstrated by the success of bioinspired polymers and composites in both clinical practice and in laboratory research. In particular, bone is the hierarchical tissue that has inspired a myriad of biomimetic materials, devices, and systems for decades. This chapter focuses on this well-developed area of biomimetic or bioinspired nanopolymers and nanocomposites for bone substitution and regeneration, especially those with high potentials for clinical applications in the near future.

4.1 Design and fabrication of bioinspired nanopolymers and nanocomposites

4.1.1 Biological synthesis of nanopolymers and nanocomposites

The natural world synthesizes a lot of polymers and composites at different scales, providing valuable principles and insights for the fabrication of nanopolymers and nanocomposites. Here, two important biological synthesis strategies, biomineralization and synthetic biology, are introduced.

Nanotechnology-Enhanced Orthopedic Materials. http://dx.doi.org/10.1016/B978-0-85709-844-3.00004-5

4.1.1.1 Biomineralization approaches

Biomineralization is the process by which living organisms produce minerals, often to harden existing tissues. In the past few decades, a significant amount of knowledge has been accumulated for natural biomineralization processes, which are extremely important in the formation of bone, tooth, and animal exoskeleton. Understanding these processes, in turn, has inspired biomimetic synthesis and fabrication of nanocomposites. There are several key players in the biomineralization process, such as specific proteins and peptides, polysaccharides and citrate molecules, and amorphous carbonates or phosphates, which trigger or facilitate the formation of minerals. Readers can refer to a brief review [3] if interested in the roles of such key players in biomineralization. Particularly for bone-like nanocomposites, a number of approaches have been developed based on the aforementioned key players.

For instance, Jee et al. developed an *in vitro* model system for producing bone-like nanocomposite via intrafibrillar mineralization, which was achieved in synthetically reconstituted collagen sponges using a polymer-induced liquid-precursor (PILP) mineralization process [4]. Through this approach, nanostructured arrangement of hydroxyapatite (HA) crystals within the individual fibrils, which mimics the structure of bone, can be produced. 3-D *in vitro* biomimetic mineralization was also achieved by utilizing nanostructured gels of self-assembling peptides containing phosphoserine residues [5]. In this strategy, natural enzyme alkaline phosphatase (ALP) and a phosphorylated, anionic nanofiber gel matrix were employed to precipitate HA nanocrystals with size, shape, and crystallographic orientation resembling natural bone mineral.

In another exemplary study, Zhong et al. reported an acid polysaccharide-based hydrogel as a 3-D matrix for biomimetic mineralization in a simulated body fluid (SBF) mineralizing system [6] (Figure 4.1). The acid polysaccharide-based hydrogels provided reactive sites for the binding of the mineral phases (due to the functionalized carboxyl moieties in maleic chitosan [7]) and, more importantly, the hydrogel played an important role in stabilizing amorphous inorganic phase at the early stage of crystallization. After mineralization, the hydrogel formed porous polysaccharide/carbonated apatite composites with tunable mineral amounts dependent on the mineralization time, as shown in Figure 4.1 [6]. Such novel porous composites can potentially serve as viable scaffolds for bone tissue engineering or as fillets for direct treatment of bone defects [3].

Recently, self-assembling peptide amphiphiles (PAs) have drawn lots of attention in biomimetic mineralization of bone-like nanocomposites [8]. PA is a molecule that consists of a peptide segment covalently bonded to a more hydrophobic segment such as alkyl tail [9] (Figure 4.2). PA molecules can self-assemble into supramolecular structures such as nanotubes or nanofibers with high aspect ratios. Some early studies have shown that these high-aspect ratio nanostructures can guide biomineralization of HA along nanofibers [10]. Moreover, the crystallographic *c*-axes of the HA crystals are aligned with the long axes of the fibers, showing an oriented epitaxial mineralization. In this process, negative charges on the PA assembly are important to establish local supersaturation of the mineral precursor, while the arrangement of the acidic groups on the fibril surface is responsible for the epitaxial mineralization [10].

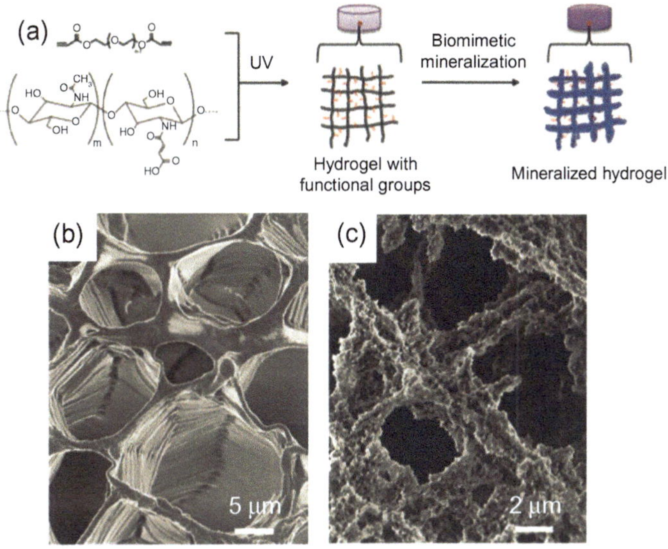

Figure 4.1 Synthesis of bone-like composites through a biomimetic strategy inspired from bone biomineralization. (a) Biomimetic composites were obtained by first cross-linking maleic chitosan with PEGDA under UV light in water to form 3-D hydrogel networks, followed by biomimetic mineralization; Images of maleic chitosan/PEGDA hydrogel (b) before and (c) after mineralization.
Reproduced with permission from Ref. [6].

4.1.1.2 Synthetic biology approach

Synthetic biology techniques are newly emerged fabrication methods for biomaterials that genetically modify the sequence and structure of protein-based biomaterials. It is expected that this new technology may provide both rational and evolutionary strategies for developing innovative biomaterials toward orthopedic applications [3]. Although only a few examples have been demonstrated, it is believed that synthetic biology may help to create ideal biomaterials for orthopedics and musculoskeletal engineering in the near future. In recent research, for example, Zhong et al. demonstrated a modular genetic strategy for the design of bioinspired hybrid nanofibers by combining mussel foot proteins (Mfps) of *Mytilus galloprovincialis* with CsgA proteins (the major subunit of *E. coli* amyloid curli fibers) [11]. CsgA contains five stacked strand–loop–strand motifs mediated by conserved residues and can self-assemble into nanofibers via a rate-limiting nucleation step followed by fibril extension. Mfp3 and Mfp5, major mussel adhesive foot proteins, have unstructured coil structures in solution and are critical to the underwater interfacial adhesion of mussels. One-pot isothermal Gibson assembly was generated to create two gene constructs (CsgA–Mfp3 and Mfp5–CsgA) independently. These hybrid nanofibers were synthesized by protein expression and refolding and, more importantly, could hierarchically self-assemble into higher-order structures that exhibit strong adhesiveness, as illustrated in Figure 4.3.

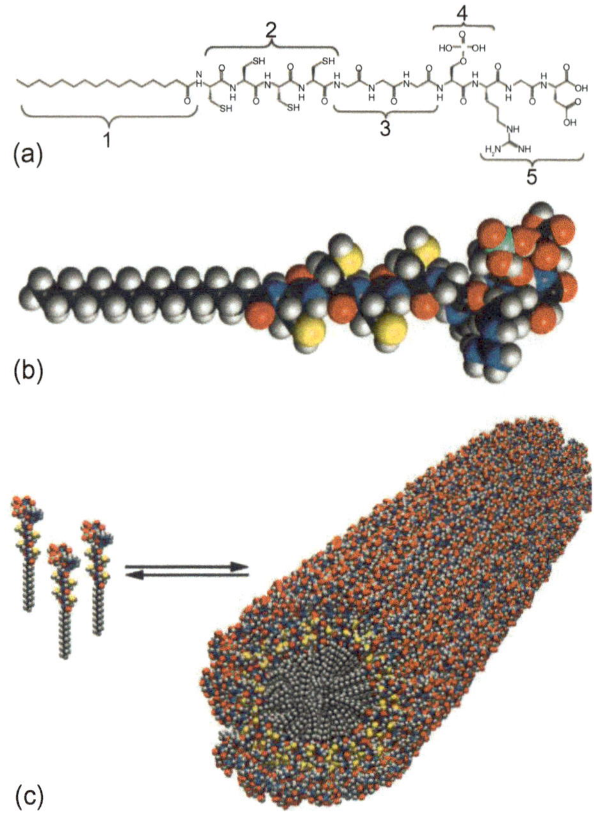

Figure 4.2 Self-assembling peptide amphiphiles (PA) used for biomimetic mineralization of HA/PA nanocomposite. (a) Chemical structure of the PA, comprising 5 regions: (1) a hydrophobic alkyl tail; (2) four cysteine residues that can form disulfide bonds to polymerize the self-assembled structure; (3) a flexible linker region of three glycine residues; (4) a single phosphorylated serine residue that was able to interact strongly with calcium ions and help direct mineralization of HA; (5) the cell adhesion ligand RGD. (b) Molecular model of one single PA molecule. (c) Schematic showing the self-assembly of PA molecules into a cylindrical micelle.
Reprinted with permission from Ref. [10].

4.1.2 Design and fabrication of multidimensional nanostructures

Design and fabrication of multidimensional nanostructures have been discussed for metals and ceramics in the previous chapters and most of the fabrication methods introduced earlier can be transferred to polymeric and composite systems. Generally, polymeric and composite nanostructures of different dimensions from 0-D (e.g., nanoparticles, nanodots) to 3-D (nanoscaffolds) can be fabricated by nanopatterning techniques, colloidal processes, electrospinning, physical and chemical vapor depositions, porogen strategy, self-assembly, and so on. These fabrication techniques discussed in the previous chapters are transferrable for preparing nanopolymers and

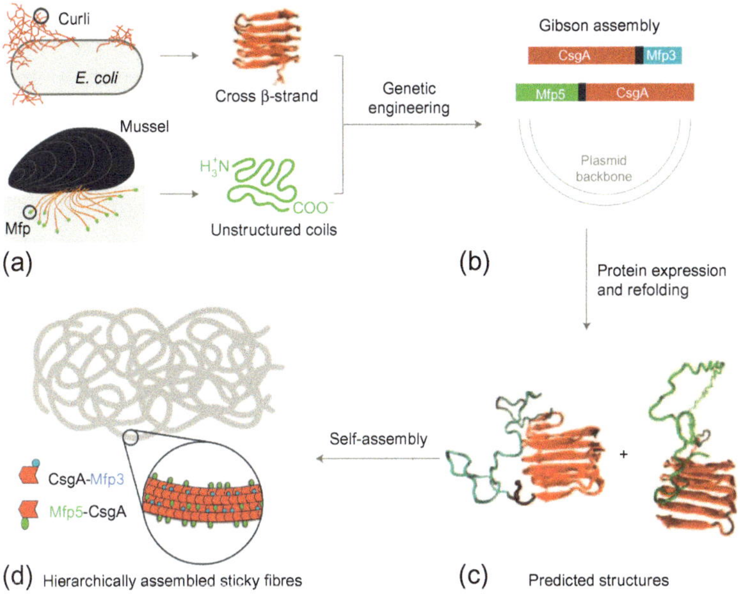

Figure 4.3 Combinatorial and modular genetic strategy for engineering self-assembling underwater adhesives. (a) Schematic of two independent natural adhesive proteins: curli from *E. coli* and Mfps from mussels. (b) The modular design of artificial adhesive materials is enabled by rationally fusing genes encoding the two natural adhesive elements shown in (a). Two gene constructs (CsgA–Mfp3 and Mfp5–CsgA) were independently created using one-pot isothermal Gibson assembly and tagged with polyhistidine residues to enable purification. (c) Schematic of predicted cross-β-strand structures for CsgA–Mfp3 and Mfp5–CsgA in solution. In both predicted structures, CsgA forms the amyloid core, with Mfp3 and Mfp5 extending from CsgA's C-terminal or N-terminal, respectively, as unstructured coil structures. (d) Because of the amyloidogenic domains, CsgA–Mfp3 and Mfp5–CsgA monomers can self-assemble into large bundles of fibrils or hierarchical networks of filaments. CsgA domains are key to fiber self-assembly by enabling fibril extension through self-polymerization and β-strand lamination by lateral stacking, with adhesive domains Mfp3 and Mfp5 exposed on the fibril surfaces. *In vitro* copolymerization of the CsgA–Mfp3 and Mfp5–CsgA monomers can lead to hierarchically coassembled structures with two different adhesive domains displayed on amyloid scaffolds, potentially recapitulating intermolecular interactions between Mfp3 and Mfp5 molecules in natural mussel adhesion systems. Enhanced underwater adhesion is expected to arise from the synergy between the high fiber surface area of curli and the adhesive residues from the Mfp domains.
Reprinted with permission from Ref. [11].

nanocomposites and, therefore, are not reiterated here. Instead, several new techniques or examples of producing such multidimensional nanostructures are introduced.

4.1.2.1 Lipid membrane-coated nanoparticles

Recently, lipid membrane-coated materials like red blood cell (RBC) membrane-coated gold nanoparticles (AuNPs) have been invented [12]. The fabrication process of RBC membrane-coated AuNPs include two steps: (1) deriving membrane vesicles from RBC by hypotonic treatment and (2) fusing the vesicles onto the surface of AuNPs by the extrusion process. Specifically, intracellular contents of RBCs are removed by hypotonic treatment and the RBC membranes are filtrated through 100 nm porous membrane to form vesicles. 70 nm citrate-stabilized AuNPs are then mixed with RBC membrane vesicles and extruded through 100 nm porous membranes several times, during which the extrusion force facilitates the fusion of RBC membrane vesicles with solid AuNPs. The resultant composite nanoparticles (i.e., RBC membrane-coated AuNPs) have diameters of 86 nm. This technique represents a group of bioderived methods to fabricate nanoparticles coated with lipid membrane or other biomolecular components.

4.1.2.2 Layer-by-layer architectures

Layer-by-layer (LbL) technique has been increasingly used for preparing 2-D nano biomaterial in the past decade due to its ability to precisely control layer structure and thickness. LbL technique is essentially based on strong interactions like electrostatic interaction between adjacent molecule layers to form assemblies. More importantly, LbL architecture provides a unique platform for creating biomimetic environment to support cell functions and tissue assembly by incorporating various materials of excellent biological and mechanical properties [13].

For example, a novel osteophilic electrostatic multilayer assembly combining HA, chitosan, recombinant human bone morphogenetic protein 2 (rhBMP-2), poly(β-amino ester) (Poly2), and poly(acrylic acid) (PAA) was developed recently by Shah et al. [13]. Figure 4.4 illustrates the design and fabrication process of the multilayer by LbL technique. In this strategy, HA/chitosan nanoparticles are incorporated into nanoscale nondegradable electrostatic multilayers that are capped with a degradable Poly2-based film incorporating physiological amounts of rhBMP-2. In this design, HA/chitosan-contained PAA films serve as osteoconductive layers and Ploly2-based hydrolytically degradable films act as osteoinductive layers that promote osteoblast differentiation by introducing controlled amounts of rhBMP-2. These two types of layers are alternatively dip-coated on glass substrates and the multilayer thickness is controlled by the number of osteoconductive HA/chitosan/PAA layers.

4.1.2.3 3-D porous scaffold by plastic compression

As mentioned in Chapter 3, one of the key factors in design and fabrication of 3-D nanoscaffolds for orthopedic applications is the appropriate porous structures. Plastic compression technique has emerged as a new method to obtain appropriated porous structures in nanopolymers and nanocomposites. More importantly, this method is friendly for incorporating proteins or biomolecules into nanostructures. Plastic compression technique, firstly reported by Brown et al. [14], has been extended to produce 3-D porous nanoscaffolds of collagen and bioactive glass-incorporated collagen [15]. In brief, this technique is composed of three main steps. First, acid solubilized collagen is neutralized in a basic solution of Dulbecco's modified Eagle's medium (DMEM) and NaOH. For producing bioactive glass/collagen nanoscaffold, nanosized bioactive glass (nBG) is added to the basic solution. Second, neutralized collagen or collagen/nBG is casted in mold and removed. Third, the casted gel is compressed at 1 kPa and the results sheet is rolled and halved to form a cylindrical, osteoid-like

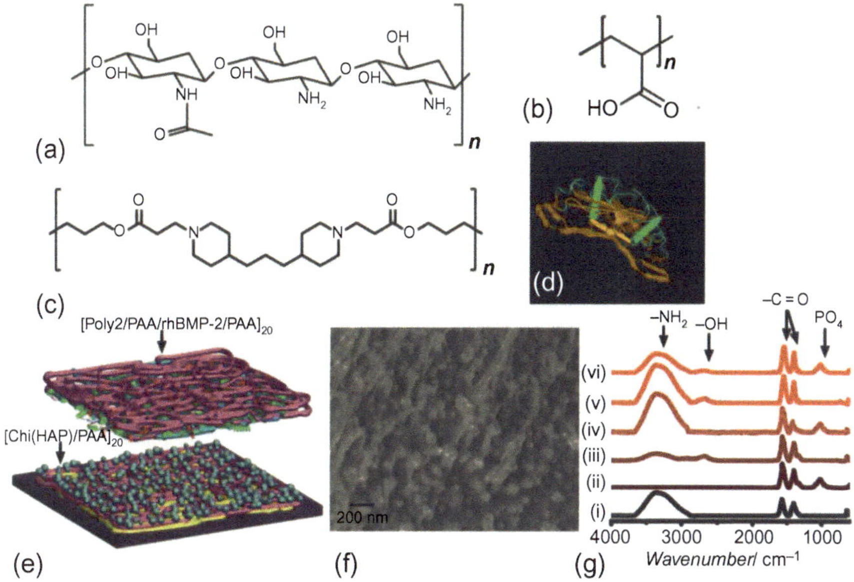

Figure 4.4 Design and fabrication of the osteophilic multilayer architecture. Components of the film (a) Chitosan (75–85% deacetylated chitin); a polycation (b) Poly(acrylic acid) (PAA); (c) A hydrolytically degradable poly(β-amino ester) (Poly2), a polycation and (d) osteoinductive recombinant human bone morphogenetic protein 2 (rhBMP-2); (e) Schematic of the modular electrostatic assembly. Osteoconductive hydroxyapatite (HAP) is complexed with chitosan and incorporated into nanoscale thick films along with poly(acrylic acid) in a bilayer architecture. A hydrolytically degradable poly(β-amino ester)-based multilayer film incorporating osteoinductive rhBMP-2 lays atop the osteoconductive layer. (f) Scanning electron micrograph of a $[Chi(HAP)/PAA]_{20}$ bilayer film with the hydroxyapatite nanoparticles, which are interwoven in the multilayer architecture and complexed to chitosan (g) IR absorbance spectra of the different components of the osteoconductive layer (i) chitosan, (ii) hydroxyapatite, (iii) poly(acrylic acid), (iv) $[Chitosan/HAP]_1$, (v) $[Chitosan/PAA]_{20}$, and (vi) $[Chi(HAP)/PAA]_{20}$.
Reprinted with permission from Ref. [13].

structure with sizes in the millimeter range. Figure 4.5 shows the morphology and microstructure of plastically pressed dense collagen (DC) and DC/nBG nanocomposite scaffolds. The plastic compression method becomes an attractive alternative to other porous nanoscaffold fabrication techniques such as freeze-drying and, moreover, it is amenable for direct and homogeneous cell incorporation during the process of collagen self-assembly [16].

4.2 Nanopolymers for orthopedic applications

4.2.1 Nanostructured polymers

The capability of nanostructured polymers in regulating the behavior of bone cells (including adhesion, growth, alignment, and elongation) and, thus, promoting bone

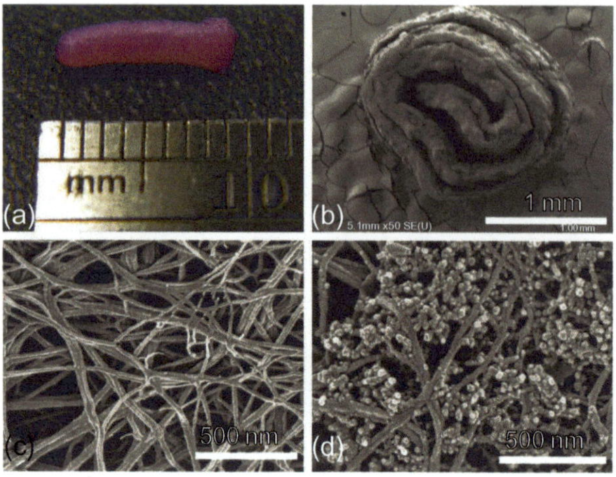

Figure 4.5 Morphological and chemical characterization of as made dense collagen (DC) and DC-nano bioactive glass (nBG) rolls. (a) Gross side view image of the dense hybrid gel roll. Reddish color is given by the presence of red phenol in DMEM used in the casting of collagen gels. (b) Top view low-magnification SEM micrograph of a hybrid roll. (c) and (d) High-magnification SEM micrographs of DC and DC-nBG gels, respectively. Collagen fibrils maintained their characteristic D-banding pattern indicating no denaturation during hybridization with nBG.
Reprinted with permission from Ref. [15].

growth and regeneration has been convincingly demonstrated. The enhancing effect of nanostructured polymers with surface feature sizes ranging from tens to hundreds of nanometers on osteoblast (bone-forming cell) responses has been well established by a great number of studies [17–20]. In light of these studies, synthetic and natural polymers (such as PLA, polyglycolic acid (PGA), PLGA, PCL, PU, gelatin, collagen, chitosan, silk, etc.) have been engineered into nanostructures with adequate structural properties to increase osteoconduction and osseointegration [18]. There are extensive examples in this area and readers can refer to plenty of publications if interested. Here, a renowned example of nano PLGA on selectively mediating the osteoblast and fibroblast (connective tissue-forming cells) functions is given. Smith et al. [21] reported that nanometer-scale rough PLGA surfaces (Figure 4.6) inhibited the adhesion of fibroblast while, interestingly, promoting osteoblast adhesion. In addition to cell adhesion density results, increased osteoblast spreading and decreased fibroblast spreading were observed on nanorough PLGA, indicating more active cell proliferation on the nanorough surfaces. Subsequent studies further confirmed that a decreased fibroblast adhesion on PLGA, PU, and PCL substrates was induced only if the surface roughness is in the range of 50–100 nm [22]. This selective mediation on different cell types renders the possibility of suppressing the activities of undesired cells such as osteoclasts (bone resorbing cells), cancer cells, or bacteria while simultaneously promoting the responses of wanted cells such as osteoblasts.

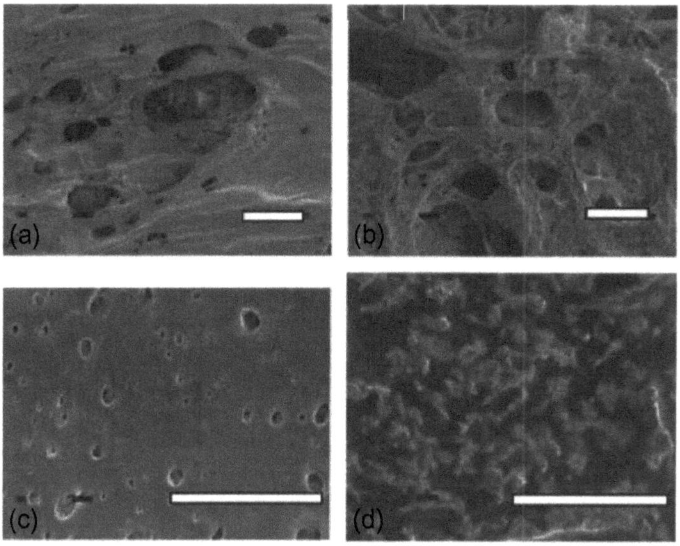

Figure 4.6 (a) SEM image of unetched PLGA scaffold showing micropores and a smooth surface (Scale bar = 100 μm). (b) SEM image of NaOH etched PLGA showing micropores and a nanorough surface (Scale bar = 100 μm). (c) SEM image of unetched PLGA showing micron-sized pores and a smooth surface (Scale bar = 10 μm). (d) SEM image of PLGA showing a nanorough surface (Scale bar = 10 μm).
Reproduced with permission from Ref. [21].

Along with the growing interest of utilizing stem cells in regenerative medicine, mediation of stem cell fate by nanobiomaterials including nanopolymers becomes a promising approach for bone repair and regeneration. Some of the earliest studies by Dalby et al. reported the control of the differentiation of human mesenchymal stem cell (MSC) via nanostructured polymethylmethacrylate (PMMA) substrates [23]. On the surface of these PMMA substrates, symmetric or disordered nanopits of 120 nm in diameter and 100 nm in depth were fabricated. The highly ordered or symmetric nanorough PMMA substrates produced very low or negligible MSC adhesion and osteoblastic differentiation, while MSCs on disordered nanosurfaces underwent osteospecific differentiation more. A pioneering study by Li and colleagues developed an electrospun PCL nanoscaffold, and MSC cultured in the nanoscaffold showed a chondrocytic phenotype differentiation [24]. Another study on nanorough PDMS and PS surfaces demonstrated that both the nanotopography and rigidity of nanopolymers influence the cell focal adhesion formation, cytoskeletal organization, and mechanical properties of human MSC [25]. Specifically, MSCs cultured on 350 nm gratings of PS and PDMS showed decreased expression of several integrin subunits (molecules participating in cell focal adhesion), compared to that on the nonpatterned polymers. MSCs on such nanopolymers also assumed elongated morphology accompanied by the alignment of actin cytoskeleton. However, the effects of nanogratings become less significant when very soft PDMS substrates are used. These results demonstrate that both topography and material rigidity of nanopolymers have significant impacts on stem cell functions.

For interaction between osteoblasts and nanopolymers, more material properties besides surface topography and rigidity have been found to affect cell activities and fate. For instance, pits-patterned PS and islands-patterned PLLA surfaces with similar surface feature sizes (~450 nm) demonstrate differentiated osteoblast adhesion results. It was found that cell adhesion is much more prominent on PLLA island surface than on PS pit surface. The study also revealed that both nano- and micron-patterned PLLA and PS surfaces could enhance cell adhesion compared to smooth PLLA and PS substrates, while cell proliferation was statistically similar on all polymer surfaces tested (i.e., nano- and micron-patterned and smooth ones) [26]. Hence, many factors including material topography, chemistry, and rigidity should be taken into account when designing high-performance nanopolymers for orthopedic applications.

4.2.2 Biomolecular nanostructures

To date, a number of nanostructures based on biomolecules including peptides or biological polymers have been devised and fabricated for musculoskeletal repair and tissue engineering applications. These nanostructures are usually in the forms of hydrogel and/or self-assembled macromolecules, which will be introduced with more details in Chapter 6. Here, a few examples on cartilage and bone regeneration are given.

For cartilage repair and bone reconstruction, biomolecular nanostructures are frequently used to encapsulate chondrocytes, cartilage or bone progenitor cells, and stem cells. Successful examples include peptides or DNA pairs that can self-assemble into high-ordered architectures. For instance, hydrogel nanoscaffolds based on self-assembling peptides KLD-12 (sequence AcN-KLDLKLDLKLDL-CNH_2) were fabricated to encapsulate chondrocytes and tested *in vitro* [27]. The 4-week *in vitro* tests revealed the promoted chondrocyte differentiation and improved synthesis of a true cartilage-like ECM rich in proteoglycans and type II collagen. Recently developed peptide nanostructures (Figure 4.7a and b), consisting of self-assembled PA (refer to

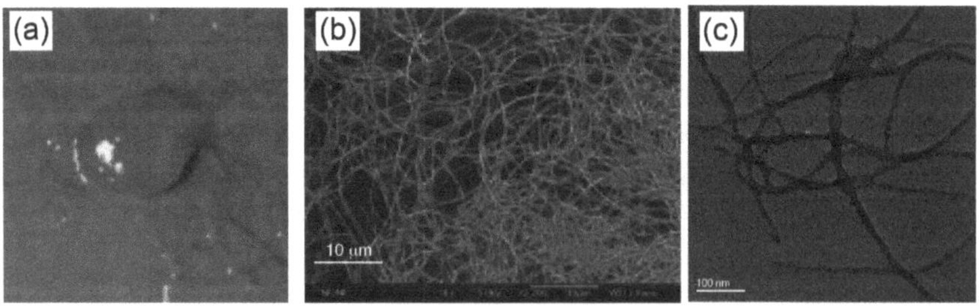

Figure 4.7 (a) Light microscopic photograph of a gel formed by adding cell culture media (DMEM) to a PA aqueous solution [29]. (b) SEM photograph of self-assembled PA nanofibers network [29]. (c) TEM image of 0.1 mg/ml helical rosette nanotubes dispersed in hydrogels [28]. Reprinted with permission from Refs. [28,29].

Figure 4.2), are shown to be effective nanoscaffolds for enhancing osteogenic response or differentiation of MSC [30]. Similarly, helical rosette nanotubes (HRN), self-assembled structures of DNA base pairs (as shown in Figure 4.7c), were fabricated into nanoscaffold and revealed osteogenic effect and capability to significantly improve osteoblast functions [31]. Besides, Shim et al. [32] fabricated nano-/microfibrous 3-D scaffolds by electrospinning chitosan nanofibers onto a predefined microfibrous mesh and such nano-/microfibrous 3-D matrix could promote chondrocyte proliferation and glycosaminoglycan synthesis *in vitro* compared to microfibrous 3-D matrix.

One great promise of biomolecular nanostructures is that they can be readily modified by other biomolecules like protein segments or peptides. For instance, PA and HRN can be further modified by arginine-glycine-aspartic acid (Arg-Gly-Arp or RGD), lysine, or KRSR (Lys-Arg-Ser-Arg) to possess even better osteogenic responses (such as enhanced ALP activity, osteocalcin synthesis, or calcium deposition) of MSC and osteoblasts than respective unmodified materials [30,31,33]. This modification will be further discussed in Chapter 6. The other promise of biomolecular nanostructures is that they can be fabricated into injectable materials or moldable soft matters that are suitable for minimally invasive surgery and/or drug release purposes. For example, injectable PA gel of nanofibrous scaffolds carrying BMP-2 was developed to promote bone regeneration through the elongated release of the osteoinductive protein for up to 24 days [34]. Similarly, injectable HRN hydrogel was created by adding RGD- or lysine-conjugated HRN (HRN-RGD-K) to polymerized 2-hydroxyethyl methacrylate (pHEMA), and *in vitro* evaluation indicated enhanced fibronectin adsorption followed by increased osteoblast adhesion on HRN-RGD-K [28].

Although nanopolymers have demonstrated a lot of promise and opportunities for orthopedic treatment and bone tissue engineering, the mechanisms of nanopolymer-mediated osteoinduction, osteoconduction, and osseointegration have not been fully understood. A general understanding is that nanostructures can enhance the adsorption of bone cell adhesive proteins (e.g., fibronectin and vitronectin) and subsequently promote bone cell attachment and osteogenic functions or inhibit functions of undesired cells. This mechanism seems valid on some types of nanopolymers. For example, it has been reported that 3-D PLA nanofibrous scaffolds enhance selective adsorption of fibronectin and vitronectin from blood, leading to increased osteoblast functions [35]. However, differentiated, even opposite, results have also been reported; therefore, this mechanism needs further verification and study.

In addition, it is important to emphasize that polymeric materials are generally soft and each type of polymer may have large variations in stiffness due to different degrees of polymerization. This differentiates them from other types of materials such as metal or ceramic, whose stiffness value is usually a constant or varies in a relatively small range. Because cell behavior can be strongly influenced by the stiffness of their microenvironment [36], cell responses on various nanopolymers are probably affected by complex interactions between material stiffness and other material properties. Therefore, the effect of nanopolymer stiffness is not negligible when considering cell-nanopolymer interaction and uncovering the mechanism of nanopolymer-mediated cell responses.

4.3 Nanocomposites for orthopedic applications

Compared to relatively soft nanopolymers, nanocomposites may receive more interest in orthopedics due to their adjustable mechanical strength and stability, which are able to match the mechanical property of bone. In this section, nanocomposites are divided into several groups based on material characteristics and corresponding strategies associated with their applications.

4.3.1 Calcium phosphate-based nanocomposites

Calcium phosphate-based nanocomposites are ideal for orthopedic applications because natural bone is essentially a nanocomposite containing polymeric substances (e.g., collagens and other ECM proteins) and calcium phosphates (e.g., HA and tricalcium phosphate (TCP)). The strategy for developing ideal orthopedic nanocomposites is optimizing the advantages of components in the composites to best mimic the hierarchical structure, microenvironment, and mechanical properties of natural bone. Along this line, a great number of calcium phosphate-based nanocomposites have been investigated to date and the nanocomposites are categorized into two slightly differentiated groups: one is obtained by combining different synthetic components of polymers and ceramics, and the other relies on the biomineralization process.

Examples of the first category include PCL/HA/gelatin nanofibrous scaffolds [37], HA/collagen nanocomposites [38], HA/chitosan nanofibers [39], PLGA/TCP nanocomposites [40], and PLGA/HA composites [41], just to name a few. These nanocomposites can properly mimic the nanometer roughness of natural bone and have been reported to promote osteoblast or MSC adhesion, ALP activity, and mineralization both *in vitro* and *in vivo*, which is well summarized in several recent books or review papers [17,42,43]. Recently, there is a research trend of conferring calcium phosphate–polymer composites with new functionalities such as directional delivery of drugs and therapeutic agents. For example, a magnetic nanocomposite scaffold of HA/collagen and iron oxide nanoparticles (200 nm) has been devised (Figure 4.8) [44]. It was proposed to use such magnetic nanocomposite scaffold, which was able to attract and carry growth factors, stem cells, or other bioagents via magnetic guiding, to treat bone defects. The nanocomposite scaffold was fabricated by dip-coating the HA/collagen biomimetic scaffold in the ferrofluid to incorporate magnetic nanoparticles. The freeze-dried nanocomposite scaffold exhibited strong magnetic effect (Figure 4.8a) and high porosity (>87 vol.%, Figure 4.8b). Preliminary *in vitro* evaluation indicated that such magnetic scaffolds supported adhesion, proliferation, and osteogensis of human bone marrow stem cells (Figure 4.8b and c). It was also found that magnetization of scaffolds did not have a significant influence on cell viability and strength of the magnetic field generated when magnetized was within the useful range for *in vivo* applications.

Another demonstrative example is the layered-scaffolds of BMP-incorporated HA nanocomposite for inducing osteogenesis and subsequent bone regeneration. A type of HA/chitosan/PAA nanocomposite, fabricated by the LbL technique introduced in

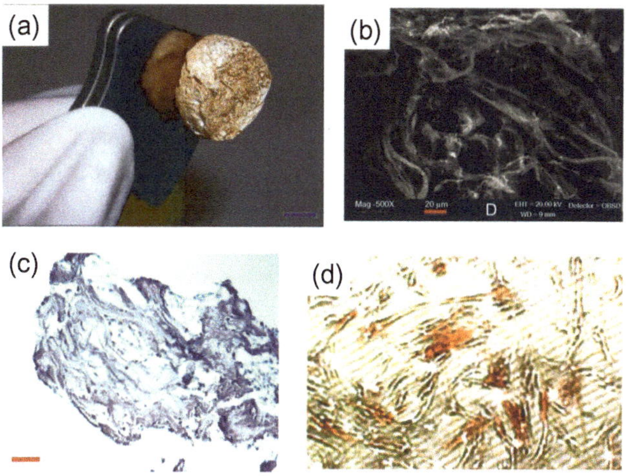

Figure 4.8 (a) Photograph showing the attraction of the porous HA/collagen/magnetic nanoparticle composite scaffold to a magnet after magnetization (Scale bar = 4 mm). (b) SEM image of the porous HA/collagen/magnetic nanoparticle composite scaffold (Scale bar = 20 μm). (c) Morphological evaluation of hBMSC cultured on the scaffold, by light microscopy and HE staining: after 10 days of cultivation, the cells have colonized the surface and most of the inner pores (Scale bar = 100 μm). (d) Alkaline phosphatase (ALP) histological staining of human bone marrow stem cells cocultivated with the scaffolds at day 15 (Magnification 100×).
Reprinted with permission from Ref. [44].

Section 4.1.2 and shown in Figure 4.4, was loaded with BMP-2 and demonstrated the ability to accelerate the regeneration of stable bone tissue [13]. The nanocomposite consisted of an osteoconductive base nanolayer and an osteoinductive top nanolayer. The former contained HA that offered a platform on which MSCs could proliferate and differentiate, while the latter could gradually release active BMP-2 over 2 weeks and induce a differentiation response on MSCs. Synergetic effects of HA and BMP-2 upregulated osteogenic markers at earlier times in MSCs, which resulted in greater calcium deposition *in vitro*. The ease of fabrication and the scalability of this hybrid multilayer film render it promising and attractive for orthopedic implant and device applications, with the goal of offering successful and lasting tissue replacement, wound healing, and joint repair [13].

Calcium phosphate-based nanocomposite can also be efficiently prepared by employing the biomineralization process, which offers a biomimetic approach similar to natural bone formation and, therefore, may provide some advantages for orthopedic applications. The first advantage is the precisely spatial and quantitative controls of calcium phosphate deposition in the organic matrix or scaffold. For example, acid polysaccharide-based hydrogel was used as a 3-D matrix immersed in a SBF mineralizing system to produce composite scaffolds for bone tissue engineering or treating bone defects [6] (Figure 4.1). Previously mentioned in Section 4.1.2, the location and amount of mineral phases can be controlled by tailoring the reactive sites in the polysaccharide and mineralization time, respectively. Secondly, the nanocomposites prepared by biomineralization may possess higher bioactivity and osteogenic properties than the nanocomposites fabricated by other approaches. For example, it has been reported that needle-like nano-HA (n-HA) mineralized in situ in an agar–gelatin cohydrogel was more favorable for the proliferation and differentiation of MC3T3-E1 osteoblasts compared to that synthesized by the traditional method [45]. The third advantage is that the size, shape, and crystallographic orientation of the calcium phosphate nanocrystals in the nanocomposites can be mediated and controlled to resemble natural bone minerals. For instance, PA peptide gels containing natural enzyme ALP and phosphorylated, anionic nanofibers were able to enzymatically mediate the harvesting of phosphate ions during biomineralization, leading to a selective control on the size, shape, and crystallographic orientation of HA. This *in vivo* success of the HA-PA composites was tested in a rat femoral defect model. The 5-mm critical-sized bone defect was filled by the nanocomposite gels and subsequent bone formation was analyzed [46]. The results showed within 4 weeks comparable bone formation to that observed in animals treated with a clinical gold standard of allogenic bone. Needless to say, these biomimetic calcium phosphate-based nanocomposites hold great potential to be applied as bioactive materials for bone tissue engineering and orthopedic treatments.

4.3.2 *Titania-based nanocomposites*

Ceramic nanoparticles like titania have also been added into polymer matrices to adjust composite surface chemistry, topography, and wettability (surface energetics) of the polymer matrix, aiming at the promotion of osteogenic responses on the material surfaces. In one of the earliest studies, nanophased titania was mixed with PLGA to produce a nanocomposite and *in vitro* evaluations revealed that osteoblast and chondrocyte adhesions were promoted on the nanocomposite compared to pure PLGA [47]. A surface wettability study of this titania/PLGA composite reported that the addition of nanosized titania to PLGA made the material surface more hydrophilic, which might enhance the initial protein interactions with substrates. A follow-up study suggested that osteoblast functions (such as adhesion collagen synthesis, ALP activity) as well as accumulation of calcium-containing mineral deposits in the extracellular matrix was probably correlated with the nanoscale roughness of the titania/PLGA composite. It also revealed that the nanocomposite with root mean square roughness close to that of natural bone might be optimal for promoting osteoblast adhesion and other cell functions including ALP activity, and mineralization compared to pure PLGA [48].

4.3.3 *Bioglass-based nanocomposites*

Similar to the strategy of titania, bioglass is also an effective nanocomponent to add in polymers for improving mechanical, biological, and osteogenic properties. Bioglass is a family of glasses usually composed of different ratios of SiO_2, Na_2O, CaO, and P_2O_5 and exhibiting high bioactivities to bind to bone and/or soft tissues. Some types of bioglass are biodegradable and can be completely resorbed within a few weeks *in vivo*. One of the most important bioglass formulations, 45S5, discovered by Larry L.

Hench in 1969, contains 45 wt.% SiO_2 and 5:1 ratio of CaO to P_2O_5. This bioglass is considered the first man-made material to bond to living tissue and has been successfully transferred to industry and clinics with FDA approvals and CE marks for dental and orthopedic products. Nanosized bioglass has also been produced and incorporated with other organic materials to make better nanocomposites for orthopedics. For example, 45S5 bioglass nanoparticles (nBG) with size distribution of 20–50 nm can be prepared by flame spray [49] and added to polymeric matrices to form nanocomposites [50].

Recently, implantable and mineralizable cell-seeded hydrogel scaffolds for bone tissue engineering were developed by hybridizing DC with osteoinductive nanosized nBG particles [15]. The fabrication technique for the DC/nBG hybrid gel scaffolds is introduced in Section 4.1.2 and the structure of the scaffolds is illustrated in Figure 4.5. The DC/nGB scaffolds were highly reactive and accelerated mineralization of carbonated HA on the scaffold was observed. After 7-day immersion in the SBF, compressive modulus of the DC/nBG scaffolds was increased 12 times. Cell culture tests showed that MC3T3-E1 preosteoblasts, three-dimensionally seeded at the point of nanocomposite self-assembly, were viable up to 28 days. More importantly, accelerated osteogenic differentiation of MC3T3-E1 preosteoblasts, which was indicated by enhanced metabolic and ALP activities, was detected in the DC/nBG scaffolds. These results indicate the high potential and suitability of the DC/nBG nanocomposites as osteoinductive cell delivery scaffolds for bone regenerative therapy.

4.3.4 Carbon-based nanocomposites

Carbon-based nanocomposites refer to a class of composites modified or reinforced by carbon nanostructures such as carbon nanotube (CNT), carbon nanofiber (CNF), and particulate nanodimond (PND). Here, the strategy of utilizing carbon nanostructures, primarily CNT and CNF, to improve osteogenic property and bioactivity of the nanocomposites is primarily discussed. The strategy of promoting mechanical properties of orthopedic implants by creating carbon-based nanocomposites will be discussed in Chapter 5.

Supported by both theoretical and experimental results, CNT and CNF have an electric current-carrying capacity over 1000 times greater than copper wires [51] and these carbon nanostructures are therefore used to improve the electrical properties of numerous biomaterials. For instance, CNT–TiN nanocomposites, composed of 12% CNTs by volume, revealed a 45% increase in electrical conductivity over TiN itself [52]. As it is well known that bone regeneration occurs under electrical conduction or simulation, CNT and CNF have been added to biomaterials in order to increase the composite conductivity and consequently promote bone growth *in vivo* or *in vitro* [53,54]. For example, when an electrical stimulatin (10 μA alternating current at 10 Hz for 6 h daily for 21 consecutive days) was applied to cell culture *in vitro*, nanocomposites of PLA and multiwalled CNTs increased osteoblast proliferation (2 days) and calcium deposition in the extracellular matrix (21 days) by 46% and 307%, respectively, and upregulated the expression of various genes (such as osteopontin, osteonectin mRNA, and osteocalcin) compared to the case without electrical stimulation [54].

Table 4.1 Carbon nanostructure-based nanocomposites for orthopedic applications

Nanocomposites	Applications
Carbon nanofiber (CNF)/polyurethane (PU) composite	Increasing osteoblast functions [56]
Poly(lactic acid) (PLA)/CNT composite	Increasing osteoblast proliferation, extracellular calcium deposition, and upregulating mRNA expression for collagen type-I under electrical stimulation [54]
Multiwall CNT/ultrahigh molecular weight polyethylene (UHMWPE) composite	Increasing cytocompatibility of osteoblast-like cells [57]
CNF/polycarbonate urethane (PCU) composite	Promoting selective adhesion and alignment of osteoblasts [58]
CNF/PLGA composite	Enhancing select osteoblast adhesion [55]

Besides the elevated conductivity, properly prepared CNT- and CNF-based nanocomposites demonstrate favorable surface topography, roughness, wettability, and/or chemistry for osteoblast responses and functions [18]. Nanophased PLGA casts of CNFs with an average diameter of 60 nm possessed a higher (by approximately 50%) degree of nanometer surface roughness compared with PLGA casts of conventional carbon fibers (average diameter >200 nm), and increased osteoblast adhesion on the nanorough CNF/PLGA composite was observed after 1 h [55]. Similar *in vitro* results have been discovered on the nanocomposites of CNT or CNF with PU, polycarbonate urethane, polyethylene, and polyethylene, which are summarized in Table 4.1.

4.4 Summary and future perspective

From a material design perspective, bioinspired nanopolymers and nanocomposites represent ideal biomaterials for bone substitution and regeneration as these material can chemically and structurally mimic natural bone, which is a hierarchical nanocomposite itself. Previous sections discuss and summarize the principles, typical examples, and recent progress of designing, fabricating, and applying nanopolymers and nanocomposites for orthopedic uses. These nanomaterials possess great potential and promise to fulfill the requirements of orthopedic materials in clinical settings, such as biocompatibility or bioactivity, osteogenicity, osteoinductivity, and osteoconductivity. In addition, these nanomaterials provide rich platforms to design and incorporate other functionalities into orthopedic implants and bone repair materials or devices.

From a future perspective, novel fabrication strategies like synthetic biology may also revolutionize orthopedic nanocomposites or nanopolymers by inducing new multi-functionalities that can regulate tissue responses and regeneration at multiple scales. These functional properties are expected to include, but are not limited to, the capabilities of (1) establishing excellent mechanical integrity with the surrounding bone or other hard tissues; (2) providing favorable 3-D environments that maintain and promote cell viability and functions; (3) possessing the abilities to deliver various growth factors, regulatory proteins, and drugs in a desired manner to stimulate tissue-specific differentiation and growth while fight infection, inflammation, and carcinogenesis; and (4) responding appropriately to surrounding physicochemical cues and stress or strain similar to those encountered in healthy bone tissues.

Another future direction is integrating nanopolymers or nanocomposites with new stem cell therapy like induced pluripotent stem cells (iPS) therapy for tissue repair and regenerations. New but large success of stem cell therapy in bone and cartilage repairs has paved a new way for treating osteoarthritis and other orthopedic diseases. Recent *in vitro* and *in vivo* studies have clearly demonstrated the potential of using nanoscaffolds to improve the efficacy and safety of stem cell therapy in the treatment of musculoskeletal defects [59]. Owing to the versatility of nanomaterials, nanopolymers and nanocomposites are expected to add new dimension to the booming area of stem cell therapy and offer more opportunities to tackle the clinical problems in orthopedic prostheses and musculoskeletal medicine.

References

[1] Miller DC, Thapa A, Haberstroh KM, Webster TJ. Endothelial and vascular smooth muscle cell function on poly (lactic-co-glycolic acid) with nano-structured surface features. Biomaterials 2004;25:53–61.
[2] Thapa A, Webster TJ, Haberstroh KM. Polymers with nano-dimensional surface features enhance bladder smooth muscle cell adhesion. J Biomed Mater Res A 2003;67:1374–83.
[3] Yang L, Zhong C. Advanced engineering and biomimetic materials for bone repair and regeneration. Front Mater Sci 2013;7:313–34.
[4] Jee SS, Kasinath RK, DiMasi E, Kim YY, Gower L. Oriented hydroxyapatite in turkey tendon mineralized via the polymer-induced liquid-precursor (PILP) process. Crystengcomm 2011;13:2077–83.
[5] Spoerke ED, Anthony SG, Stupp SI. Enzyme directed templating of artificial bone mineral. Adv Mater 2009;21:425–30.
[6] Zhong C, Chu CC. Biomimetic mineralization of acid polysaccharide-based hydrogels: towards porous 3-dimensional bone-like biocomposites. J Mater Chem 2012;22:6080–7.
[7] Zhong C, Wu J, Reinhart-King CA, Chu CC. Synthesis, characterization and cytotoxicity of photo-crosslinked maleic chitosan-polyethylene glycol diacrylate hybrid hydrogels. Acta Biomater 2010;6:3908–18.
[8] Palmer LC, Newcomb CJ, Kaltz SR, Spoerke ED, Stupp SI. Biomimetic systems for hydroxyapatite mineralization inspired by bone and enamel. Chem Rev 2008;108:4754–83.
[9] Palmer LC, Stupp SI. Molecular self-assembly into one-dimensional nanostructures. Acc Chem Res 2008;41:1674–84.
[10] Hartgerink JD, Beniash E, Stupp SI. Self-assembly and mineralization of peptide-amphiphile nanofibers. Science 2001;294:1684–8.
[11] Zhong C, Gurry T, Cheng AA, Downey J, Deng Z, Stultz CM, et al. Strong underwater adhesives made by self-assembling multi-protein nanofibres. Nat Nanotechnol 2014;9:858–66.
[12] Gao W, Hu CÄJ, Fang RH, Luk BT, Su J, Zhang L. Surface functionalization of gold nanoparticles with red blood cell membranes. Adv Mater 2013;25:3549–53.

[13] Shah NJ, Hong J, Hyder M, Hammond PT. Osteophilic multilayer coatings for accelerated bone tissue growth. Adv Mater 2012;24:1445–50.
[14] Brown RA, Wiseman M, Chuo C, Cheema U, Nazhat SN. Ultrarapid engineering of biomimetic materials and tissues: fabrication of nano-and microstructures by plastic compression. Adv Funct Mater 2005;15:1762–70.
[15] Marelli B, Ghezzi CE, Mohn D, Stark WJ, Barralet JE, Boccaccini AR, et al. Accelerated mineralization of dense collagen-nano bioactive glass hybrid gels increases scaffold stiffness and regulates osteoblastic function. Biomaterials 2011;32:8915–26.
[16] Ghezzi CE, Muja N, Marelli B, Nazhat SN. Real time responses of fibroblasts to plastically compressed fibrillar collagen hydrogels. Biomaterials 2011;32:4761–72.
[17] Zhang LJ, Webster TJ. Nanotechnology and nanomaterials: promises for improved tissue regeneration. Nano Today 2009;4:66–80.
[18] Yang L, Zhang LJ, Webster TJ. Nanobiomaterials: state of the art and future trends. Adv Eng Mater 2011;13:B197–217.
[19] Liu H, Webster TJ. Nanomedicine for implants: a review of studies and necessary experimental tools. Biomaterials 2006;28:354–69.
[20] Balasundaram G. Nanomaterials for better orthopedics. In: Webster TJ, editor. Nanotechnology for the regeneration of hard and soft tissues. Hackensack, NJ; London: World Scientific; 2007. p. 53–78.
[21] Smith LL, Niziolek PJ, Haberstroh KM, Nauman EA, Webster TJ. Decreased fibroblast and increased osteoblast adhesion on nanostructured NaOH-etched PLGA scaffolds. Int J Nanomed 2007;2:383.
[22] Vance RJ, Miller DC, Thapa A, Haberstroh KM, Webster TJ. Decreased fibroblast cell density on chemically degraded poly-lactic-co-glycolic acid, polyurethane, and polycaprolactone. Biomaterials 2004;25:2095–103.
[23] Dalby MJ, Gadegaard N, Tare R, Andar A, Riehle MO, Herzyk P, et al. The control of human mesenchymal cell differentiation using nanoscale symmetry and disorder. Nat Mater 2007;6:997–1003.
[24] Li WJ, Tuli R, Huang XX, Laquerriere P, Tuan RS. Multilineage differentiation of human mesenchymal stem cells in a three-dimensional nanofibrous scaffold. Biomaterials 2005;26:5158–66.
[25] Yim EK, Darling EM, Kulangara K, Guilak F, Leong KW. Nanotopography-induced changes in focal adhesions, cytoskeletal organization, and mechanical properties of human mesenchymal stem cells. Biomaterials 2010;31:1299–306.
[26] Wan Y, Wang Y, Liu Z, Qu X, Han B, Bei J, et al. Adhesion and proliferation of OCT-1 osteoblast-like cells on micro-and nano-scale topography structured poly (L-lactide). Biomaterials 2005;26:4453–9.
[27] Kisiday JD, Kurz B, DiMicco MA, Grodzinsky AJ. Evaluation of medium supplemented with insulin-transferrin-selenium for culture of primary bovine calf chondrocytes in three-dimensional hydrogel scaffolds. Tissue Eng 2005;11:141–51.
[28] Zhang L, Rakotondradany F, Myles AJ, Fenniri H, Webster TJ. Arginine-glycine-aspartic acid modified rosette nanotube-hydrogel composites for bone tissue engineering. Biomaterials 2009;30:1309–20.
[29] Hosseinkhani H, Hosseinkhani M, Kobayashi H. Design of tissue-engineered nanoscaffold through self-assembly of peptide amphiphile. J Bioact Compat Polym 2006;21:277–96.
[30] Hosseinkhani H, Hosseinkhani M, Tian F, Kobayashi H, Tabata Y. Osteogenic differentiation of mesenchymal stem cells in self-assembled peptide-amphiphile nanofibers. Biomaterials 2006;27:4079–86.
[31] Zhang LJ, Ramsaywack S, Fenniri H, Webster TJ. Enhanced osteoblast adhesion on self-assembled nanostructured hydrogel scaffolds. Tissue Eng Pt A 2008;14:1353–64.

[32] Shim IK, Suh WH, Lee SY, Lee SH, Heo SJ, Lee MC, et al. Chitosan nano-/microfibrous double-layered membrane with rolled-up three-dimensional structures for chondrocyte cultivation. J Biomed Mater Res A 2009;90A:595–602.
[33] Zhang LJ, Chen YP, Rodriguez J, Fenniri H, Webster TJ. Biomimetic helical rosette nanotubes and nanocrystalline hydroxyapatite coatings on titanium for improving orthopedic implants. Int J Nanomed 2008;3:323–33.
[34] Hosseinkhani H, Hosseinkhani M, Khademhosseini A, Kobayashi H. Bone regeneration through controlled release of bone morphogenetic protein-2 from 3-D tissue engineered nano-scaffold. J Control Release 2007;117:380–6.
[35] Woo KM, Chen VJ, Ma PX. Nano-fibrous scaffolding architecture selectively enhances protein adsorption contributing to cell attachment. J Biomed Mater Res A 2003;67A:531–7.
[36] Discher DE, Janmey P, Wang YL. Tissue cells feel and respond to the stiffness of their substrate. Science 2005;310:1139–43.
[37] Venugopal JR, Low S, Choon AT, Kumar AB, Ramakrishna S. Nanobioengineered electrospun composite nanofibers and osteoblasts for bone regeneration. Artif Organs 2008;32:388–97.
[38] Kikuchi M, Itoh S, Ichinose S, Shinomiya K, Tanaka J. Self-organization mechanism in a bone-like hydroxyapatite/collagen nanocomposite synthesized in vitro and its biological reaction in vivo. Biomaterials 2001;22:1705–11.
[39] Zhang Y, Venugopal JR, El-Turki A, Ramakrishna S, Su B, Lim CT. Electrospun biomimetic nanocomposite nanofibers of hydroxyapatite/chitosan for bone tissue engineering. Biomaterials 2008;29:4314–22.
[40] Schneider OD, Weber F, Brunner TJ, Loher S, Ehrbar M, Schmidlin PR, et al. In vivo and in vitro evaluation of flexible, cottonwool-like nanocomposites as bone substitute material for complex defects. Acta Biomater 2009;5:1775–84.
[41] Jose MV, Thomas V, Johnson KT, Dean DR, Nyairo E. Aligned PLGA/HA nanofibrous nanocomposite scaffolds for bone tissue engineering. Acta Biomater 2009;5:305–15.
[42] Zhou H, Lee J. Nanoscale hydroxyapatite particles for bone tissue engineering. Acta Biomater 2011;7:2769–81.
[43] Liu H, Webster TJ. Bioinspired nanocomposites for orthopedic applications. In: Webster TJ, editor. Nanotechnology for the regeneration of hard and soft tissues. Singapore: World Scientific Publishing; 2007. p. 1–51.
[44] Bock N, Riminucci A, Dionigi C, Russo A, Tampieri A, Landi E, et al. A novel route in bone tissue engineering: magnetic biomimetic scaffolds. Acta Biomater 2010;6:786–96.
[45] Deng Y, Wang H, Zhang L, Li Y, Wei S. In situ synthesis and in vitro biocompatibility of needle-like nano-hydroxyapatite in agar-gelatin co-hydrogel. Mater Lett 2013;104:8–12.
[46] Mata A, Geng YB, Henrikson KJ, Aparicio C, Stock SR, Satcher RL, et al. Bone regeneration mediated by biomimetic mineralization of a nanofiber matrix. Biomaterials 2010;31:6004–12.
[47] Kay S, Thapa A, Haberstroh KM, Webster TJ. Nanostructured polymer/nanophase ceramic composites enhance osteoblast and chondrocyte adhesion. Tissue Eng 2002;8:753–61.
[48] Liu H, Slamovich EB, Webster TJ. Increased osteoblast functions among nanophase titania/poly(lactide-co-glycolide) composites of the highest nanometer surface roughness. J Biomed Mater Res A 2006;78A:798–807.
[49] Brunner TJ, Grass RN, Stark WJ. Glass and bioglass nanopowders by flame synthesis. Chem Commun 2006;13:1384–6.
[50] Boccaccini AR, Erol M, Stark WJ, Mohn D, Hong Z, JoF Mano. Polymer/bioactive glass nanocomposites for biomedical applications: a review. Compos Sci Technol 2010;70:1764–76.
[51] Avouris P, Appenzeller J, Martel R, Wind SJ. Carbon nanotube electronics. Proc IEEE 2003;91:1772–84.

[52] Jiang L, Gao L. Carbon nanotubes/metal nitride composites: a new class of nanocomposites with enhanced electrical properties. J Mater Chem 2005;15:260–6.
[53] Harrison BS, Atala A. Carbon nanotube applications for tissue engineering. Biomaterials 2007;28:344–53.
[54] Supronowicz PR, Ajayan PM, Ullmann KR, Arulanandam BP, Metzger DW, Bizios R. Novel current-conducting composite substrates for exposing osteoblasts to alternating current stimulation. J Biomed Mater Res 2002;59:499–506.
[55] Price RL, Ellison K, Haberstroh KM, Webster TJ. Nanometer surface roughness increases select osteoblast adhesion on carbon nanofiber compacts. J Biomed Mater Res A 2004;70A:129–38.
[56] Webster TJ, Waid MC, McKenzie JL, Price RL, Ejiofor JU. Nano-biotechnology: carbon nanofibres as improved neural and orthopaedic implants. Nanotechnology 2004;15:48–54.
[57] Reis J, Kanagaraj S, Fonseca A, Mathew MT, Capela-Silva F, Potes J, et al. In vitro studies of multiwalled carbon nanotube/ultrahigh molecular weight polyethylene nanocomposites with osteoblast-like MG63 cells. Braz J Med Biol Res 2010;43:476–82.
[58] Khang D, Sato M, Price RL, Ribbe AE, Webster TJ. Selective adhesion and mineral deposition by osteoblasts on carbon nanofiber patterns. Int J Nanomedicine 2006;1:65–72.
[59] Brown PT, Handorf AM, Jeon WB, Li WJ. Stem cell-based tissue engineering approaches for musculoskeletal regeneration. Curr Pharm Des 2013;19:3429–45.

Carbon nanostructures: new materials for orthopedic applications

5

The discovery and emergence of carbon nanostructures have impacted and reshaped many aspects of nanotechnology, and have boosted developments in physics, electronics, optics, mechanics, biology and medicine since the 1990s. By far, the discovered or synthetic carbon nanostructures include 0-D structures (i.e., fullerenes, particulate diamond and carbon black), 1-D structures (i.e., carbon nanotubes or nanofibers, and diamond nanorods), 2-D structures (i.e., graphene, graphite sheets, and diamond nanopletelets), and 3-D structures (i.e., nanocrystalline diamond scaffolds/coatings, nanostructured diamond-like carbon scaffolds/coatings, and fullerite), as illustrated in Figure 5.1 [1–3]. Due to their extraordinary properties and differentiated carbon hybridization status (e.g., sp^2, sp^3 hybridization), many carbon nanostructures have attracted steadily increasing attention in different fields of biology and medicine. For example, fullerene and carbon nanotubes have been studied for therapeutic and pharmaceutical purposes soon after their discoveries in the 1990s [1]. However, the history of carbon nanostructures for orthopedic applications is relatively short. In this chapter, progress and the state of the art in this new area will be covered, including important aspects such as fabrication techniques, attractive properties, and promise of the carbon nanostructures as new orthopedic materials.

5.1 Carbon nanostructures and fabrication methods

Depending on the degrees of carbon bond hybridization in carbon materials (i.e., sp^2, sp^3 hybridization), various fabrication techniques have been developed to prepare different carbon nanostructures. These techniques are introduced individually for each category of carbon nanostructures.

5.1.1 Carbon nanotube (CNT) and carbon nanofiber (CNF)

CNTs are cylindrical carbon nanostructures that are either single-walled CNTs (SWCNTs) consisting of a graphene layer rolled into a perfect cylinder (Figure 5.2a), or multi-walled CNTs (MWCNTs) composed of several concentric cylindrical graphene layers (Figure 5.2b). Similarly, CNFs are cylindrical structures (not as perfect as CNTs) with graphene layers stacked into bundles, cones, or plates (Figure 5.2c) [3]. Chemical bonding of CNT or CNF is composed of entirely sp^2 bonds, similar to those of graphene and graphite. For the laboratory and industrial production of CNT and CNF, chemical vapor deposition (CVD) and arc-discharge deposition are

Nanotechnology-Enhanced Orthopedic Materials. http://dx.doi.org/10.1016/B978-0-85709-844-3.00005-7

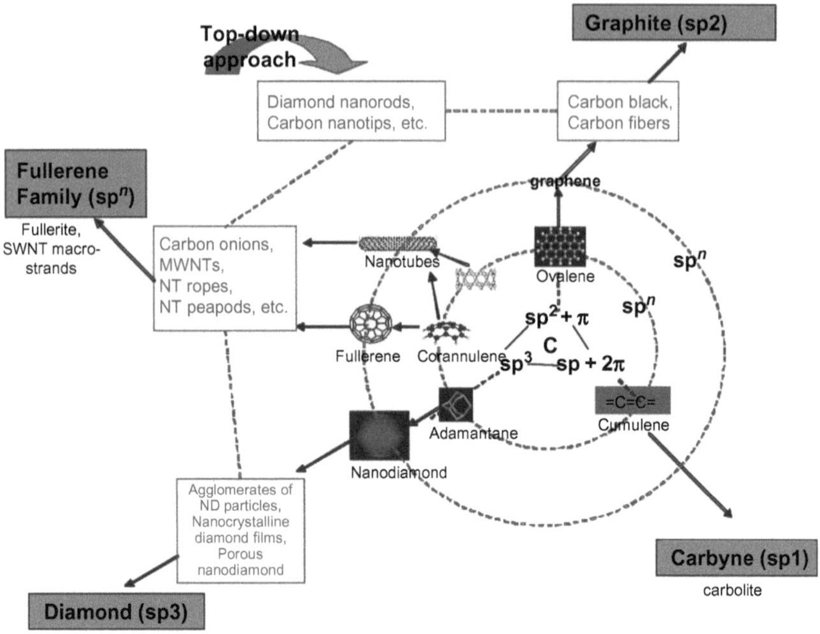

Figure 5.1 The family of carbon nanostructures. The mark "sp^n" indicates intermediate carbon forms with a non-integer degree of carbon bond hybridization.
Adapted from Ref. [1] with permission.

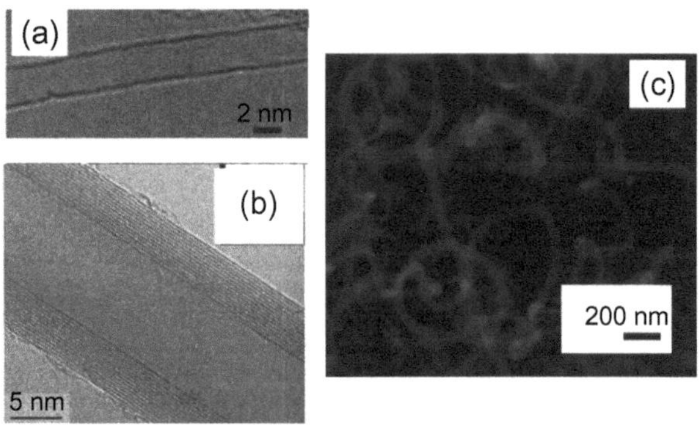

Figure 5.2 Transmission electron microscopy (TEM) images of (a) Single-walled carbon nanotube (SWCNT), (b) multi-walled carbon nanotube (MWCNT), and (c) carbon nanofibers (CNFs). Figure 2a and b were retrieved from Refs. [4,5] with permissions, respectively.

two most widely used methods (Table 5.1). These two methods can be also used to fabricate other carbon nanostructures like nanodiamonds and graphene, which will be discussed later.

5.1.1.1 Chemical vapor deposition

There are two major types of CVD techniques depending on power input methods. Thermal CVD involves heating a hydrocarbon to a high temperature usually in the

Table 5.1 **Fabrication methods of CNTs and CNF [3]**

Carbon nanostructures	Fabrication methods
MWCNT	• Chemical vapor deposition (CVD) • Arc-discharge deposition • Laser vaporization deposition • Ion beam assisted deposition
SWCNT	• CVD • Arc-discharge deposition • Laser vaporization deposition • High-pressure CO conversion (HiPCO)
CNF	• CVD • Arc-discharge deposition

presence of a catalyst (e.g., metallic nanoparticles of Ni, Co, or Fe) which causes the hydrocarbon to decompose and then deposit onto a substrate (e.g., silicon, glass, or alumina). This method has been successfully adapted in the production of CNT and CNF using different carbon feedstock [6]. The length, diameter, alignment and quality (often evaluated by the amounts of defects and amorphous carbon) of CNT or CNF can be controlled by the chemical composition and textural properties of the catalyst material(s) used in the CVD process [7].

In addition, CNTs and CNFs can be synthesized using plasma enhanced CVD (PECVD) in which the hydrocarbon gas is in an ionized state over the catalyst(s). Figure 5.3 shows the schematic and photograph of a microwave-assisted PECVD system. One advantage of PECVD over thermal CVD is that carbon nanomaterial deposition can occur at a lower substrate temperature (as low as room temperature in some cases). Quality and structures of carbon nanomaterials (e.g., length, diameter, and alignment of CNTs or CNFs) can also be controlled by altering carbon feedstock, deposition parameters or catalyst properties [8]. Both of the CVD methods are capable of producing carbon nanostructures in large quantities [7,8].

5.1.1.2 Arc-discharge deposition

The arc-discharge deposition is a method in which a graphitic rod, acting as a negative cathode and a positive anode when placed a few millimeters apart, are evaporated using a high current to form carbon products which deposit around the chamber walls or over the cathode substrate [9]. CNTs and CNFs can be synthesized in the arc-discharge AC (alternating current) or DC (direct current) systems, but DC provides higher yields of CNTs (mostly MWCNTs) which are deposited on the cathode [6]. For producing SWCNTs, metal catalysts (such as Ni, Fe, and Co) need to be coated on the graphite anode while pure graphite is used as cathode [10,11]. Arc-discharge deposition is capable of synthesizing CNTs into large yield quantity (gram quantities) [12] and in lengths greater than 40 μm [13].

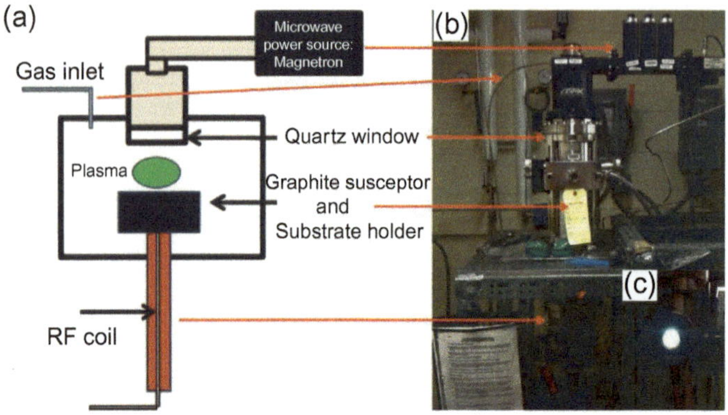

Figure 5.3 (a) Schematic showing the setup of the microwave-assisted PECVD system for growing CNT, CNF, or NCD on a substrate, (b) photograph of the PECVD system, and (c) photograph of the $Ar/CH_4/H_2$ plasma as viewed from observation window.

5.1.1.3 High-pressure CO conversion

High-pressure CO conversion (HiPCO) is a new method for the bulk production of SWCNTs under high-pressure, high-temperature flowing CO on catalytic clusters of Fe. Fe catalyst is formed *in situ* by thermal decomposition of iron pentacarbonyl (i.e., $Fe(CO)_5$) which is delivered intact within a cold CO flow and then rapidly mixed with hot CO in the reaction zone. Upon heating, the $Fe(CO)_5$ decomposes into atoms that condense into larger clusters. SWCNTs nucleate and grow on these particles in the gas phase via CO disproportionation: $CO + CO(\text{catalyzed}) \rightarrow CO_2 + C(\text{SWCNT})$ [14,15].

5.1.2 Fullerene

Fullerenes (e.g., C_{60}, C_{70}, C_{84}, buckyballs) are usually referred to spherical molecular crystals composed entirely of carbon atoms and the atoms are arranged into pentagonal, hexagonal, or sometimes heptagonal rings (Figure 5.1). In fullerenes, carbon bond hybridization can be described as sp^n, where n is not integer but $2 < n < 3$, indicating intermediate carbon forms. As illustrated in Figure 5.1, fullerene is structurally similar to SWCNT so they are arguably categorized as the same types of carbon structures (i.e., fullerene family).

There are several methods to synthesize fullerenes and the frequently used ones are arc-discharge deposition and laser vaporization deposition techniques. The arc-discharge deposition method has been covered above and fullerenes are deposited on the cathode substrate by evaporating carbon from graphite rods when applying a voltage in the presence of an inert gas [16]. The use of inert gas instead of hydrogen-containing gas is the essential difference between CNT and fullerene production by arc-discharge method. In addition, the arc-discharge method can be optimized to produce fullerenes in large quantities.

5.1.2.1 Laser vaporization deposition

The laser vaporization (sometimes laser ablation) method has also been widely used for synthesizing fullerenes since 1985 [1]. This method typically involves vaporizing a target material (e.g., graphite or polycarbonate) by a focused laser beam (typically a yttrium aluminum garnet (YAG) or CO_2 laser). The vapor is then cooled down in a noble gas buffer (e.g., Ar) and deposited on a collector [6]. Comparing to the arc-discharge method, the laser vaporization deposition has relatively lower yield in producing fullerene.

5.1.3 Nanodiamonds and nanostructured diamond-like carbon

Nanodiamonds indicate synthetic diamond materials that have sizes or feature sizes (e.g., grain sizes) between 1 and 100 nm, generally including two forms: particulate and thin film. Because the nanodiamond thin film is actually composed of nanosized columnar crystals or grains, the material is also known as nanocrystalline diamond. Due to its close relationship and similarity to nanodiamonds, nanostructured diamond-like carbon has also attracted considerable attention. In the following sections, these carbon nanostructures are discussed separately.

5.1.3.1 Particulate nanodiamond (PND)

Essentially, PND is dispersed, nanoscale diamond clusters or crystals with a size range of a few nanometers to tens of nanometers. Figure 5.4a shows the morphology of PND under scanning electron microscope (SEM). One of the most frequently used techniques for manufacturing PND is detonation, which has already been used on industrial scale. In 1988, PND (sizes 4–7 nm) formed in detonation process was made known to the world [18] and the synthesis methods of PND by detonation have been developed since the 1990s [19,20]. The principles of detonation synthesis include the detonation of an explosive in oxygen-deficient environment so that some elements in the explosive (such as C, H, and N) cannot be oxidized; during detonation process, free carbon atoms are released and then coagulate and crystallize to form nanosized diamond under the high temperature and high pressure produced by the detonation reaction [19,21]. PND usually comprises up to 75% of the denotation soot, so the

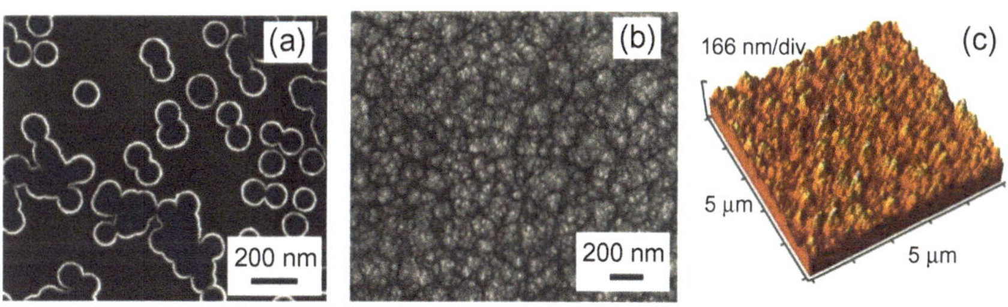

Figure 5.4 SEM images of (a) particulate nanodiamond (PND) and (b) nanodiamond thin film or nanocrystalline diamond (NCD), (c) AFM image of NCD.
Reprinted with permission from Ref. [17].

soot need to be purified to obtain the PND. The primary PND crystallites produced by detonation have a size range of 2–10 nm but are very likely to agglomerate into aggregates of tens or hundreds of nanometers [19].

5.1.3.2 Nanocrystalline diamond (NCD)

NCD or sometimes ultra-NCD (UNCD, grains sizes <10 nm) is continuous thin film deposited on a substrate material and its thickness ranges from tens of nanometers to a few microns. In comparison to graphene that has a thickness of one or a few atomic layers (see the section below), NCD is sometimes categorized as 3-D carbon nanostructure. Figure 5.4b and c show the NCD thin film under SEM and atomic force microscope (AFM), respectively. To fabricate NCD thin films, CVD, glow discharge, and carbon sputtering techniques are widely accepted by both laboratory and industry. The fabrication condition of each technique is listed in Table 5.2.

The most widely used technique is again CVD, including microwave-assisted PECVD, hot filament CVD and DC arc CVD [22,24,29]. These techniques share similar deposition mechanisms but differentiate in power generation methods for sustaining the chemical vapor reactions. A typical microwave-assisted PECVD system including a schematic of its setup is shown in Figure 5.3. For preparing NCD, the plasma of $Ar/CH_4/H_2$ gas mixture is usually used, as shown in Figure 5.3c. The PECVD system, however, is versatile for preparing NCD, CNT, and other carbon structures.

A typical CVD process of growing NCD includes substrate polishing and cleaning, seeding PND on the substrate (creating sites to initiate growth), and film deposition from chemical reaction of precursors. The substrate materials include Si, SiO_2, Si_3N_4, metals, glass, and so on. The regular precursors are CH_4/H_2 mixtures with addition of inert gases, Ar or N_2 [29]. The parameters of CVD deposition (deposition temperature, chamber pressure, bias voltage, deposition time, etc.) play critical but complicated roles in the resultant NCD quality (e.g., sp^2/sp^3 contents), grain sizes and film thickness as well as NCD properties.

Table 5.2 CVD techniques and conditions for fabricating NCD films

Techniques	Typical growth conditions	References
Microwave plasma CVD (MPCVD)	$CH_4/H_2/Ar$ or CH_4/H_2 mixtures	[22,23]
DC arc CVD	CH_4/H_2/50% Ar mixture	[24]
Hot filament CVD	CH_4/H_2 or 1–4% acetone/H_2 mixtures	[22,25]
Glow discharge	CH_4/H_2 mixture	[26]
Carbon sputtering	High pressure magnetron sputtering in Ar/0–10% H_2	[27]
Electron-cyclotron resonance assisted CVD	CH_4 at low pressure with bias voltage applied	[28]

5.1.3.3 Nanostructured diamond-like carbon (NanoDLC)

Diamond-like carbon (DLC) is hard, amorphous carbon film with a significant fraction of sp^3-hybridized carbon atoms and a significant amount of hydrogen. DLC films can be fully amorphous or contain diamond crystallites [29]. NanoDLC usually refers to the DLC films possessing nanoscale structures and roughness or nanocrystallite inclusions (e.g., PND or diamond clusters). Fabrication techniques of DLC, including CVD, arc, and discharge deposition methods which have been reviewed above, can be used to prepare NanoDLC by modifying deposition conditions or parameters [30].

Besides, ion beam-assisted deposition has also been employed for growing NanoDLC. This deposition method usually involves a process which bombards a substrate with high energy ions of a hydrocarbon (such as CH_4) in a vacuum such that hydrogen ions are liberated upon dissociation of the hydrocarbon molecules [31]. During a direct ion beam deposition process, a carbon-containing ion beam of controlled composition, energy, and flux is directed onto the substrate [32]. Direct ion beam deposition is mainly used to prepare DLC or NanoDLC with high adhesion, high purity, and high optical transparency since the ion beam could remove weakly bound carbon, physisorbed residual gases, and impurities [33–36]. However, ion beam-assisted deposition has also been used to synthesize nanostructured carbon films with well-organized graphene layers (up to ~8 nm) [37], nanographite [38], and MWCNTs [39].

5.1.4 Graphene

Graphene is actually the single 2-D layer of graphite with one-atom-thick, sp^2-bonded honeycomb carbon lattice. This material was first isolated from graphite by exfoliation with adhesive tape in 2004 and fabrication methods of graphene have expanded quickly since then. Graphene and its derivatives are such new carbon nanostructures that their research and applications in orthopedics are still limited. Here, their fabrication methods are briefly overviewed and the readers can refer to literature [40] for more details.

Basically, there are two strategies for producing graphene: (1) cleaving multi-layer graphite into single layers and (2) growing it epitaxially by depositing one atomic layer of carbon onto a substrate material. The former usually uses mechanical methods like adhesive tape cleavage and sonication to split the single-atom-layer carbon from carbon. Whereas, the latter strategy is often achieved by CVD techniques. For example, large-area graphene can be grown on Cu or Ni foils by CVD using CH_4 and H_2 at temperatures up to 1000 °C [41].

Among the fabrication methods introduced above, one can find that CVD is probably the most versatile approach to fabricate carbon nanostructures from CNT/CNF to NCD to graphene. CVD techniques have also been adopted by industry for many years and this adoption paves a broad road for scaling up the production of carbon nanostructures for various applications.

5.2 Carbon nanotube, carbon nanofiber, and fullerene for orthopedic applications

It is well known that CNTs and CNFs possess exceptional electrical and mechanical properties, such as extreme high conductivity and strength, low weight-to-strength ratios, and so on. They also have nanoscale dimensions in diameters or lengths similar to crystalline minerals and biomolecules found in bone. These advantages have attracted immense attention from orthopedic researchers and open lots of possibilities for their use in orthopedic applications. To date, the potential orthopedic applications that would benefit from the use of CNTs/CNFs focus on three aspects: mechanical reinforcement, improvement in biological properties, and new functions such as biosensing and drug delivery. In addition, due to structural similarity between fullerene and CNT, the application of fullerenes in orthopedics is also introduced here.

5.2.1 Orthopedic implants reinforced by CNT and CNF

In the past decade, CNTs and CNFs have been actively studied for improving mechanical properties (strength, facture toughness, fatigue life, and wear-resistance) of orthopedic implant materials. For this purpose, these carbon nanostructures have been studied as reinforcements for ceramics, polymers and hydrogels, as summarized in Table 5.3. For example, both CNTs and CNFs reveal robust reinforcing effects on the mechanical properties of ceramic compositions. MWCNTs distributed in hydroxyapatite (HA) plasma sprayed coatings showed 56% increase in fracture toughness (reached to 0.61 MPa $m^{1/2}$) compared to plain HA coating (0.39 MPa $m^{1/2}$) [49].

Table 5.3 Applications of CNT and CNF in improving mechanical properties of orthopedic implants

Nanostructure	Application in improving mechanical properties of implants
MWCNT	• Enhancing fatigue life and mechanical properties of bone cement [42–44] • Increasing compression strength, electrical conductivity and bioactivity of bioglass scaffolds [45,46] • Improving bending strength and bending modulus of polymer/MWCNT nanocomposites [47,48] • Enhancing the flexural strength, fracture toughness and fatigue performance of ceramic/MWCNT or polymer/MWCNT composites [49–51] • Improving wear-resistance of polymer/MWCNT composites [52]
SWCNT	• Increasing compressive modulus, flexural modulus, compressive offset yield strength and flexural strength of polymer/SWCNT composites [53,54]
CNF	• Improving the fracture toughness of HA/CNF mixture [55]

A powder mixture of CNF/HA showed similar bending strength values to cortical bone (~90 MPa) and the fracture toughness values for CNF/HA were around 1.6 times higher than those obtained for monolithic HA ceramics [55]. Recently, Meng et al. fabricated a new hydroxyapatite–zirconia–carbon nanotube (HA-ZrO_2-CNT) ceramic composite to significantly enhance the flexural strength (increased by ~126%) and fracture toughness (increased by ~124%) of the HA ceramic [50]. The addition of MWCNTs to highly porous 45S5 Bioglass® foam scaffolds (one of the most widely used bioactive glasses for bone tissue engineering) also improved the compression strength as well as electrical conductivity and bioactivity of the bioglass [45,46].

The reinforcing effects of CNTs have also been reported in other composite formulations. Marrs et al. found that the addition of MWCNTs to bone cement or methyl methacrylate–styrene copolymers (MMA-co-Sty; the main component of commercial bone cement) could enhance the fatigue life (evaluated by Weibull mean, N_{wm}) of the materials up to sixfold (N_{wm}=863216) depending on the weight percentage of added MWCNTs [42,51]. A similar enhancement by adding MWCNTs has been discovered in poly(methyl methacrylate) (PMMA)-based bone cements as well [43,44]. Besides bone cements, a recent study showed that a nanocomposite with 0.1 wt% functionalized SWCNTs by 4-*tert* butylbenzenediazonium tetrafluoroborate and biodegradable poly(propylene fumarate) (PPF) had threefold increases in both compressive modulus and flexural modulus and twofold increases in both compressive offset yield strength and flexural strength compared to pure PPF networks [53]. Another study found that the bending strength and bending modulus of chitosan/MWCNTs rods (100/0.5, W/W) increased by 34.3% and 7.3% (increased to 130.7 MPa and 4.4 GPa), respectively, compared to those of pure chitosan [47].

In addition, CNTs are effective additives to improve wear-resistance of orthopedic composite materials. Reis et al. [52] reported that a composite of ultrahigh molecular weight polyethylene (UHMWPE) and MWCNTs possessed superior wear behavior to UHMWPE only, suggesting the potential to reduce the revision of hip arthroplasty caused by wear. Lahiri et al. showed that the addition of 4 wt% CNTs to plasma-sprayed HA coatings reduced the volume of wear debris generation by 80% due to improved elastic modulus and fracture toughness of the composite [56]. The same study also revealed that HA/CNT wear debris containing loose CNTs did not increase cytotoxic effects (evaluated by cell viability) to macrophages and osteoblasts compared to HA.

5.2.2 CNT and CNF for bone repair and regeneration

The first study on CNF supporting osteoblast (bone forming cell) functions that are important for orthopedic implant applications was reported in 2002 [57]. The study reported increased osteoblast proliferation, intracellular protein synthesis, alkaline phosphatase activity, and calcium-containing mineral deposition on nanometer-diameter CNF compared to conventional micron-diameter carbon fibers. Since then, CNTs and CNFs have been widely studied for promoting the biocompatibility and biological properties of implants, regenerating bone and cartilage and reducing inflammation or other adverse effects, which are summarized in Table 5.4.

Table 5.4 **Effects of CNT and CNF on improving biological responses to orthopedic implants and their applications**

Nanostructure	Applications in improving biological responses
MWCNT	• Improving biocompatibility and bone cell adhesion, attachment and/or proliferation on collagen/MWCNT scaffolds [58], PLGA/HA/MWCNT scaffolds [59], PMMA/HA/MWCNT scaffolds [60], PLLA/MWCNT composites [61], and UHMWPE/MWCNT composites [52] • Little inflammatory response to the host and promoting new bone tissue formed in MWCNT/polymer composites [62,63]
SWCNT	• Promoting HA formation on modified SWCNT films [64] • Increasing bone cell alkaline phosphatase activity and matrix mineralization *in vitro* [65] • Improving *in vivo* bone formation and bone repair capacities of nanocomposites [65] and hydrogel system [66] • Increasing the development and calcium deposition of marrow stromal cells (MSCs) [67–69] • Reducing inflammatory cell density and increasing connective tissue organization on polymer/SWCNT scaffolds [70]
CNF	• Directing osteoblast alignment, enhancing cell adhesion, calcium phosphate mineral deposition on CNF/polymer composites [71,72]

Among studies to date, both CNTs or CNFs and CNT- or CNF-based composites have demonstrated great promise for tissue regeneration purposes. For example, Armentano et al. investigated *in vitro* biomineralization process of human alveolar bone-derived cells on carboxylated and fluorinated SWCNT films. The results showed that cellular responses (including secreting a mineralized extracellular matrix and consequent mineralization) were promoted on SWCNT films and the promotion of tissue growth depended on the degree of functionalization of the SWCNTs [64]. Green et al. reported that under photoacoustic stimulation, at day 16, marrow stromal cells (MSCs) incubated with SWCNTs at 10 μg/mL concentrations showed up to 97% greater calcium content than those that did not contain SWCNTs in tissue culture plates [67]. Besides SWCNTs, MWCNT blocks made from fluorinated MWCNTs of optimal machinability and strength (average bending strengths of 102.2 MPa and a bending modulus of 15.4 GPa) also revealed acceptable biocompatibility *in vivo*; implantation tests in the subcutaneous tissue of rats indicated less of an inflammatory response to MWCNT than PMMA [48]. More importantly, Webster's group also reported that microscale CNF linear patterns (30 μm wide lines) on polycarbonate urethane (PCU) may be used to control spatial osteoblast (bone-forming cells) attachment [73]. Results showed the selective adhesion and alignment of osteoblasts on CNF patterns and greater attraction forces between fibronectin (a protein that mediates osteoblast adhesion) and CNF (compared with PCU) patterns. Moreover, the study also showed that the directed osteoblast adhesion on CNF patterns translated to enhanced calcium phosphate mineral deposition along linear patterns of CNFs on PCU, similar to the anisotropic pattern of mineralization in long bones in the body.

For bone or cartilage prosthetics and tissue engineering, CNTs and CNFs have been widely incorporated into a number of tissue engineering polymers including PMMA [43,44], PCU [72,73], poly(L-lactide) (PLLA) [61], polycarbosilane (PCS), and poly (lactic-co-glycolic acid) (PLGA) [59]. These nanocomposites revealed improved cyto-/tissue-compatibility properties, enhanced tissue regeneration, and reduced inflammatory responses compared to pristine polymers. For example, histological investigations of MWCNT/PCS composites implanted in rat subcutaneous tissue and femurs showed little inflammatory responses and newly formed bone tissue after 1 and 4 weeks of implantation [62]. Another 12-week study of SWCNTs/PPF scaffolds implanted in rabbit femoral condyles and in subcutaneous pockets revealed a threefold greater amount of bone tissue ingrowth, reduced inflammation and increased connective tissue organization compared to polymer scaffold control [70]. Recent progress has also demonstrated that composites with functionalized CNTs possessed better biocompatibility properties and tissue regeneration properties than unmodified CNTs. A study of comparing pristine SWCNTs and carboxylated SWCNTs (SWCNTs–COOH) in PLLA polymer matrices showed that PLLA/SWCNTs–COOH nanocomposites had higher cell adhesion and proliferation than PLLA or PLLA/SWCNTs nanocomposites [61]. Similarly, a new COOH-functionalized MWCNTs composite integrated with PMMA and HA showed more favorable biocompatibility and osseointegration properties *in vitro* and *in vivo*, and more new bone ingrowth to the composite in animal studies [74].

The biological and physical mechanisms of enhanced bone cell and tissue responses to CNTs or CNFs and their composites are among the most popular research topics, although a thorough understanding of these mechanisms has not been well established. CNTs and CNFs are highly conductive and have a strong affinity to bind, immobilize, and trap proteins, which may have an important impact on cellular signal pathways, cell adhesion and subsequent cell activities [48]. When constructed into 3-D or large-scale scaffolds, factors such as nanotopography, nanoscale roughness and surface free energy of CNTs and CNFs also play important roles in regulating biological responses. For example, Firkowska et al. studied osteoblast responses to the nanotopography of cross-linked MWCNTs, showing that cells sense and react to nano topographies with excellent spreading and metabolic activity. They also found that the osteoblasts adhered to a cavity-like topography on the MWCNT-based matrices expressed an increased average Young's modulus [75]. Similarly, nanotopographies of CNF compacts and PLGA casts molded out of the CNF compacts both demonstrated greater osteoblast adhesion than micron topographies of conventional carbon fibers and PLGA [76]. In addition, Price et al. demonstrated that osteoblast adhesion was promoted on composites of PCU and CNF compared to the composites of PCU and conventional micron dimension carbon fibers [72]. They also showed that osteoblast adhesion increased whereas smooth muscle cell, fibroblast, and chondrocyte adhesion decreased with an increase in CNF surface free energy. The study of Tutak et al. also supported that the initial attachment and proliferation of osteoblasts on a substrate are strongly dependent on the energy and roughness of the surface [68].

5.2.3 CNT-based bio-sensing in orthopedic implants

Recently, a novel approach of sensing new bone growth on orthopedic implants has been proposed based on the promising electrical properties of CNTs. Sirivisoot et al. studied cyclic voltammograms of several types of CNT-modified titanium substrates and reported that MWCNTs grown out of nanotubular structures of anodized titanium (a-Ti) generated superior redox signals on such MWCNT/a-Ti substrates than unmodified Ti. Their subsequent study designed a bone growth sensor by coating a conductive, biodegradable, polymer layer (polypyrrole and PLGA) onto MWCNT/a-Ti. Therefore, when new bone grows onto the implant, the conductive polymer layer degrades and/or is replaced by newly formed bone and, consequently, undergoes a change in conductivity that can be measured by the parallel MWCNTs grown out of the pores of a-Ti. Preliminary *in vitro* results suggested that osteoblast functions (adhesion and proliferation) on such a biosensor were not significantly compromised when compared to currently-used titanium, yet, retained the ability to potentially measure new bone growth juxtaposed to an implant. These results imply a significant promise of using MWCNT/a-Ti to sense *in situ* new bone growth next to orthopedic implants and perhaps monitor other events (such as infection and/or harmful scar tissue formation) to improve the current clinical diagnosis of orthopedic implants [77–79].

5.2.4 Orthopedic medical application of fullerenes

The potentials of fullerene (C_{60}, not water soluble) and its derivatives (maybe water soluble) for supporting bone cell growth and treating arthritis have been studied. Grausova et al. [80] investigated osteoblast-like MG63 cell adhesion and growth on continuous or micropatterned fullerene films deposited on glass coverslips. Results indicated cell adhesion and proliferation activities on all kinds of fullerene films were similar to those on tissue culture polystyrene (TCPS) dishes, and region-selective cell adhesion and growth were observed on the grooves of the micropatterns. Similar results were reported on binary fullerene/titanium composites (C_{60}:Ti = 1:1) [81]. Gonzalez et al. [82] designed a fullerene-based material, $C_{60}(OH)_{16}$AMBP (4,4-bisphosphono-2-(polyhydroxyl-1,2-dihydro-1,2-methanofullerene(60)-61-carboxamido)butyric acid), to specifically target HA in bone tissues, showing the promise of fullerene derivatives for treating bone-related diseases. Fullerene functionalized with diphosphonate groups were expected to possess the same ability to selectively target bone tissues [83]. Yudoh et al. conducted a series of studies on using water-soluble fullerene to treat arthritis [84–86]. Their studies used water-soluble fullerene as an oxygen free radical scavenger to reduce articular cartilage degeneration and inhibit osteoclast differentiation and bone destruction. However, although fullerene has exhibited the strong potentials in orthopedic medical field, it is important to be aware of its possible toxicity problems. There are many studies conducted to reveal the possible health risks of fullerene [87–89], which will be covered in Chapter 9.

5.3 Particulate nanodiamond, nanocrystalline diamond and nanostructured diamond-like carbon for orthopedic medical applications

5.3.1 Particulate nanodiamond (PND)

Because of its small sizes and extraordinary mechanical, electrical, and optical properties, PND has been added to biocompatible polymers to increase material hardness and Young's modulus and to provide additional functionalities such as imaging, sensing, and tissue growth abilities. Recently, a multifunctional bone scaffold material has been developed utilizing biodegradable PLLA and octadecylamine-functionalized PND (PND-ODA), as shown in Figure 5.5a [90,91]. Addition of 10 wt% of PND-ODA resulted in more than 200% increase in Young's modulus, 800% increase in hardness, 280% increase in the strain at failure, and 310% increase in fracture energy compared to PLLA alone. These mechanical properties are close to that of the human cortical bone. Moreover, PND-ODA provided bright fluorescence for medical imaging (Figure 5.5b) and options for drug loading and delivery. These results indicated PND/polymer nanocomposite scaffolds are potentially useful for orthopedic prostheses. However, due to its structural nature, PND may not be a perfect anti-wear coating material and the application of PND for bone implant coatings is very limited.

More importantly, PND has demonstrated its promising osteogenic or osteoconductive properties that are attractive for bone regenerative engineering. *In vitro* cell tests on the PND-ODA/PLLA mentioned above reveal that the nanocomposite as a matrix supported murine osteoblast (7F2) cell growth and differentiation for up to 1 week, and enhanced biomineralization capability of forming bonelike apatite on the scaffolds in simulated body fluid (SBF). There are also recent studies showing that PND can stimulate deposition of calcium phosphate or HA on TiCu alloys or stainless steel when immersed in a mixture of SBF and PND suspension [92,93]. This

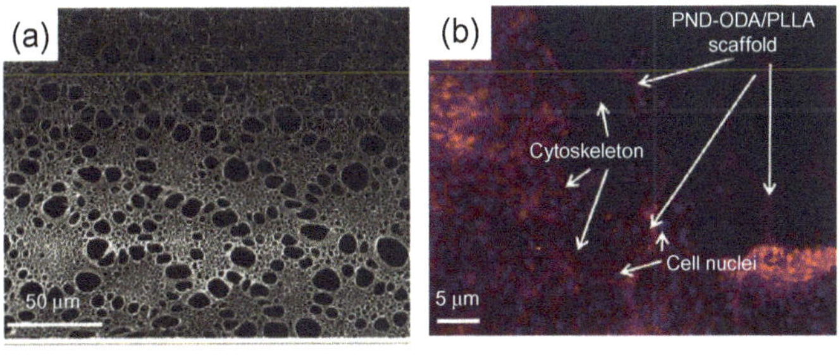

Figure 5.5 (a) The SEM image of the porous PND-ODA/PLLA nanocomposite and (b) the clear fluorescence of the nanocomposite scaffold seen in the background of osteoblasts after 3 days.
Modified from Ref. [90] with permissions.

demonstrates the osteoconductive property of PND and its possible application in fabricating bioactive coatings on metallic orthopedic implants.

PND also exhibited antioxidant, anti-inflammatory, anti-carcinogenic, and anti-allergic properties, which are also wanted in orthopedic treatments [94,95]. The structural advantages of particles actually imparted PND with novel properties or functions that are not possessed by 2-D or 3-D carbon nanostructures. For example, aqueous dispersible PND hydrogels with a diameter of 2–8 nm were able to assemble into a multilayer nanofilm with ability to capture anti-inflammatory molecules, implying a potential application in creating anti-inflammatory implant surfaces [96]. In addition, PND can be readily functionalized with biomolecules and this versatility is important for developing drug delivery or molecular imaging functions in orthopedic implants [29,95]. For instance, PND has been reported to conjugate with tissue growing hormones [97], insulin [98], fluorescent molecules, lysozyme, vaccines, and drug molecules [95].

5.3.2 Nanocrystalline diamond (NCD)

Compared to other coating materials, NCD films possess extremely good tribological behaviors because it retains most of the mechanical advantages of micron-sized or polycrystalline diamond (e.g., super hard surface) and it has very smooth topography (roughness is in the nanoscale) [22]. Thanks to the development of CVD techniques, NCD films can be deposited onto many substrates (e.g., Ti and its alloys, stainless steel, and ceramic) with satisfactory adhesion strengths, and thus can be coated on current orthopedic implants mainly made of Ti and its alloys as well as stainless steel [99–101]. Thus, NCD films have been mainly studied for implant coating proposes to provide wear- and corrosion-resistance [3].

In addition to the well-know superior mechanical and anti-wear properties, recent studies on the osteogenetic ability of NCD films have achieved tremendous progress toward improving orthopedic applications like orthopedic implant coatings. Early studies focused on the cytocompatibility properties of NCD films and the investigation of basic functions of bone cells on NCD [102–104]. For example, osteoblast (MC3T3 cell line) attachment assays on NCD were compared to platinum and silicon after 20 hours, and the results showed that NCD had superior characteristics (including cell number, total cell area, and cell spreading) over other materials [103]. In addition, equivalent or better osteoblast adhesion, proliferation and differentiation and bone marrow cells on NCD compared to TCPS were reported [104,105]. Since 2008, relevant studies entered a new phase to investigate the possible material properties that affect bone cell functions and how to manipulate these properties to achieve desired cell responses. Yang et al. [106] reported, for the first time, that both short- and long-term functions of osteoblast could be significantly changed by just altering NCD crystallinity or grain size (i.e., surface chemistry and wettability remain the same), suggesting the NCD nanotopography would be an adjustable factor to control osteoblast responses (Figure 5.6). In their other study, grain sizes and surface topographies of diamond were dramatically altered by adjusting H_2 concentration in the PECVD process, and corresponding variations in osteoblast adhesion and proliferation were shown [107]. Other groups also reported similar *in vitro* results [108,109]. These

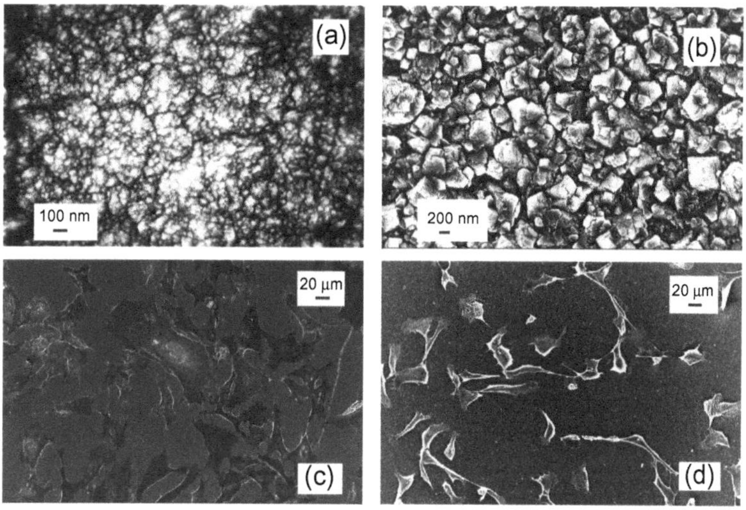

Figure 5.6 Altering NCD crystallinity or grain size to control osteoblast adhesion and proliferation *in vitro*: SEM images of (a) NCD and (b) submicron crystalline diamond (SMCD) films; enhanced osteoblast density, proliferation and spreading on (c) NCD films compared to (d) SMCD films.
Modified from Ref. [106] with permissions.

experimental results are also supported and partly verified by computational simulations which concluded that NCD nanotopography and nanoscale roughness (root mean square roughness usually <25 nm) could enhance osteoblast adhesion compared to microcrystalline diamond (MCD) [110]. Recently, an *in vivo* test of implanting NCD and MCD into the femoral diaphysis of rats confirmed previous *in vitro* findings, showing more intense tissue growth on NCD than MCD after 4 and 8 weeks [111].

Based on these results, a strategy of using NCD as orthopedic implant coatings was proposed (as illustrated in Figure 5.7): NCD with small grain sizes (usually <80–100 nm) can be applied to non-articulating surfaces where bone tissue bonding is highly desirable (Region A in Figure 5.7), whereas NCD with large grain sizes (>80–100 nm or submicrons) can be used on the articulating surfaces where bone growth is unnecessary but wear resistance is crucial (Region B in Figure 5.7).

Besides changing crystallinity and topography, manipulating surface chemistry of NCD is another approach to enhance its osteogenic properties. Kloss et al. [112] terminated NCD with oxygen by thermal treatment and further functionalize the surface with bone morphological protein-2 (BMP-2). *In vivo* tests revealed enhanced osteoconduction on the treated NCD compared to titanium and untreated NCD, and *in vitro* tests of mesenchymal stromal cells (MSC) yielded strong osteogenic expression on BMP-2 adsorbed NCD. Another study examined MSC adhesion and survival on hydrogen, oxygen and fluorine-terminated UNCD, and the results indicated the best MSC responses on hydrogen-terminated UNCD [113]. Further study on MSC osteoblastic differentiation revealed hydrogen-terminated UNCD had similar or better MSC adhesion, proliferation and differentiation compared to titanium alloys and CoCrMo.

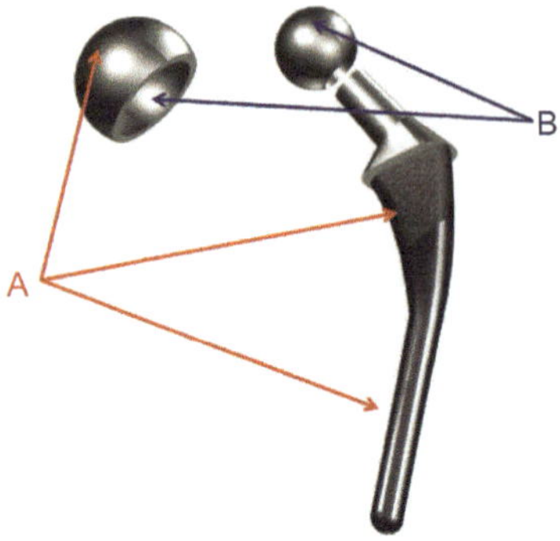

Figure 5.7 Schematic of the strategy using varied diamond coatings in different parts of an orthopedic implant. Region A represents non-articulating surfaces that require good bone growth while region B represents articulating surface where bone growth is not necessary.

However, other studies demonstrated significantly better adhesion and/or osteogenic responses of cells on oxygen-terminated NCD than hydrogen or fluorine terminated NCD [114–116]. In a recent study, plasma treatments were also developed to directly modify NCD surfaces with amine terminations for better bone growth [116]. A later study revealed that better osteoblast and fibroblast adhesion (4–12 hours) and higher cell spreading area occurred on amine-terminated NCD than oxygen- and hydrogen-terminated NCD. Moreover, altering NCD surface energy [117] and doping the NCD surface by boron [118] also demonstrated increases in osteogenic responses (adhesion, proliferation and differentiation) of osteoblasts.

Furthermore, NCD has also demonstrated a promise for reducing bacterial infection, which is highly desired for orthopedic implant materials because invasive surgery is inevitably associated with infection. A preliminary study compared the number of *Escherichia coli* colonized on NCD, stainless steel and titanium, and showed that NCD has the highest bacterial resistance [119]. This possible anti-infection effect is highly desirable for orthopedic implants and, thus, should be further studied.

5.3.3 Nanostructured diamond-like carbon (NanoDLC)

Because of its excellent mechanical and tribological properties (though not as good as diamond), biological properties and comparatively low costs, conventional DLC films intended for orthopedic applications have been studied since the 1990s and developments in this field can be found in a review [120]. Here, only advances of NanoDLC are highlighted. Recently, pertinent tribological properties of NanoDLC toward orthopedic medical applications have been reported [121,122]. NanoDLC film produced by femtosecond laser pulses with further coating of MoS_2 revealed a minimal friction coefficient of 0.02 compared to that of 0.08 on conventional and uncoated DLC (tested on WC-Co ball) [122]. However, there are not many studies on biological properties of nanostructured DLC film. A study of biocompatibility of DLC nanofilms indicated good bone cell attachment compare to other DLC films [123]. Another study shows that calcium-oxygen modified NanoDLC reduced hardness and elastic modulus but increased cell adherence compared to unmodified DLC [124]. Based on this result and many other studies on conventional DLC films [120], one can speculate that modification (e.g., doping) of NanoDLC can effectively improve its biocompatibility properties. In summary, although relevant researches are still in their infancy and further biological studies are still missing, NanoDLC can be a promising candidate to fulfill the desirables of high wear-resistance and osteoconductivity.

5.4 Promises of graphene and its derivatives as new orthopedic materials

As a newly emerged material received a lot of attention, graphene has properties such as high conductivity, high strengthen and considerable flexibility which may be interested to orthopedic engineers. Fan et al. [125] fabricated graphene-reinforced chitosan composites and evaluated their mechanical and cytocompatibility properties. Their study showed that addition of 0.1–0.3 wt% graphene increased the elastic modulus of chitosan by over 100% and the composites were biocompatible to L929 cells. Czarnecki et al. [126] studied osteoblast attachment, proliferation, and elongation on carbon fibers. They found that graphene orientations at the cell–carbon interface guided osteoblast orientation or alignment. These results provided valuable information for design and fabrication of graphene-based orthopedic materials.

5.5 Summary and future directions

Clearly, carbon nanostructures have demonstrated tremendous promise for orthopedic medical applications to date, but a number of critical problems need further investigation and solid solutions before eventual clinical practice. In fact, there are not any commercial orthopedic products based on carbon nanostructures so far. This is probably due to several unsolved problems as described below.

First, investigation on biological responses of carbon nanostructures is still incomplete. On one hand, majority of the studies so far have been only *in vitro* tests of relevant cells, but *in vivo* animal tests or ultimate human tests are missing. On the other hand, cell or tissue responses based on variations in structures or properties of carbon nanostructures (e.g., graphene orientations, grain boundary chemistry, nanotopography, surface energies and surface functional groups) remain unclear. Further understanding of these issues will enable the rational design of carbon nanostructure-based orthopedic materials.

Toxicity of carbon nanostructures has not been disclosed or understood. While toxicity of fullerenes and CNT are still under debate, relevant studies on new carbon nanomaterials like NCD and graphene have not commenced. Toxicity of materials has a determinative impact on their ultimate application and use, and thus understanding toxicity of carbon nanostructures (both during manufacturing and use) is as imperative as medical application study.

Lastly, as mentioned in the previous sections, most of the carbon nanostructures (e.g., CNT, CNF, NCD films, NanoDLC) have been devised as orthopedic implant coatings. These applications require strong and durable bonding between carbon nanostructures and substrate materials. Although a number of tribological studies have focused on the adhesion strength and the stability of the coatings, more systematic investigations which can provide a further understanding and better techniques to achieve desirable adhesion strength have not been completed. On top of all that, extensive testing of the coating durability under actual usage environments, especially the tests for long term time periods, are necessary.

References

[1] Shenderova OA, Zhirnov VV, Brenner DW. Carbon nanostructures. Crit Rev Solid State Mater Sci 2002;27:227–356.
[2] Hu YH, Shenderova OA, Hu Z, Padgett CW, Brenner DW. Carbon nanostructures for advanced composites. Rep Prog Phys 2006;69:1847–95.
[3] Yang L, Zhang L, Webster TJ. Carbon nanostructures for orthopedic medical applications. Nanomedicine 2011;6:1231–44.
[4] Sun Z, Barron A. TEM imaging of carbon nanomaterials. Connexions Web site; 2010.
[5] Endo M. Application to the composite material of carbon fiber. Endo Laboratory Web site; 2011.
[6] Yahya N, Koziol K, Boskovic BO. Synthesis of carbon nanostructures by CVD method. Carbon and oxide nanostructures. Berlin, Heidelberg: Springer; 2011, p. 23–49.
[7] Cassell AM, Raymakers JA, Kong J, Dai HJ. Large scale CVD synthesis of single-walled carbon nanotubes. J Phys Chem B 1999;103:6484–92.
[8] Ren ZF, Huang ZP, Xu JW, Wang JH, Bush P, Siegal MP, et al. Synthesis of large arrays of well-aligned carbon nanotubes on glass. Science 1998;282:1105–7.
[9] Baddour C, Briens C. Carbon nanotube synthesis: a review. Int J Chem React Eng 2005;3:1542–6580.
[10] Iijima S, Ichihashi T. Single-shell carbon nanotubes of 1-nm diameter. Nature 1993;363:603–5.
[11] Bethune DS, Klang CH, de Vries MS, Gorman G, Savoy R, Vazquez J, et al. Cobalt-catalysed growth of carbon nanotubes with single-atomic-layer walls. Nature 1993;363:605–7.
[12] Ebbesen TW, Ajayan PM. Large-scale synthesis of carbon nanotubes. Nature 1992;358:220–2.
[13] Colbert DT, Zhang J, Mcclure SM, Nikolaev P, Chen Z, Hafner JH, et al. Growth and sintering of fullerene nanotubes. Science 1994;266:1218–22.
[14] Nikolaev P. Gas-phase production of single-walled carbon nanotubes from carbon monoxide: a review of the HiPco process. J Nanosci Nanotechnol 2004;4:307–16.
[15] Bronikowski MJ, Willis PA, Colbert DT, Smith KA, Smalley RE. Gas-phase production of carbon single-walled nanotubes from carbon monoxide via the HiPco process: a parametric study. J Vac Sci Technol A 2001;19:1800–5.
[16] Merchan-Merchan W, Saveliev AV, Kennedy L, Jimenez WC. Combustion synthesis of carbon nanotubes and related nanostructures. Prog Energ Combust Sci 2010;36:696–727.
[17] Yang L, Webster T. Small and bright: nanodiamonds for tissue repair, drug delivery, and biodetection. IEEE Pulse 2014;5:34–9.

[18] Greiner NR, Phillips D, Johnson J, Volk F. Diamonds in detonation soot. Nature 1988;333:440–2.

[19] Dolmatov VY. Detonation synthesis ultradispersed diamonds: properties and applications. Russ Chem Rev 2001;70:607.

[20] Chen P, Ding Y, Chen Q, Huang F, Yun S. Spherical nanometer-sized diamond obtained from detonation. Diam Relat Mater 2000;9:1722–5.

[21] Chen Q, Yun S. Nano-sized diamond obtained from explosive detonation and its application. Mater Res Bull 2000;35:1915–9.

[22] Gruen DM. Nanocrystalline diamond films. Annu Rev Mater Sci 1999;29:211–59.

[23] Lee JC, Hong BY, Messier R, Collins RW. Nucleation and bulk film growth kinetics of nanocrystalline diamond prepared by microwave plasma-enhanced chemical vapor deposition on silicon substrates. Appl Phys Lett 1996;69:1716–8.

[24] Konov VI, Smolin AA, Ralchenko VG, Pimenov SM, Obraztsova ED, Loubnin EN, et al. Dc-Arc plasma deposition of smooth nanocrystalline diamond films. Diam Relat Mater 1995;4:1073–8.

[25] Wang T, Xin HW, Zhang ZM, Dai YB, Shen HS. The fabrication of nanocrystalline diamond films using hot filament CVD. Diam Relat Mater 2004;13:6–13.

[26] Heiman A, Gouzman I, Christiansen SH, Strunk HP, Comtet G, Hellner L, et al. Evolution and properties of nanodiamond films deposited by direct current glow discharge. J Appl Phys 2001;89:2622–30.

[27] Kundu SN, Basu M, Maity AB, Chaudhuri S, Pal AK. Nanocrystalline diamond films deposited by high pressure sputtering of vitreous carbon. Mater Lett 1997;31:303–9.

[28] Zarrabian M, FourchesCoulon N, Turban G, Marhic C, Lancin M. Observation of nanocrystalline diamond in diamondlike carbon films deposited at room temperature in electron cyclotron resonance plasma. Appl Phys Lett 1997;70:2535–7.

[29] Kharisov BI, Kharissova OV, Chavez-Guerrero L. Synthesis techniques, properties, and applications of nanodiamonds. Synth React Inorg Met 2010;40:84–101.

[30] Mousinho AP, Mansano RD, Salvadori MC. Nanostructured diamond-like carbon films characterization. J Alloy Compd 2010;495:620–4.

[31] Devries RC. Synthesis of diamond under metastable conditions. Annu Rev Mater Sci 1987;17:161–87.

[32] Kimock FM, Knapp BJ. Commercial applications of ion-beam deposited diamond-like carbon (DLC) coatings. Surf Coat Technol 1993;56:273–9.

[33] Lau WM, Bello I, Feng X, Huang LJ, Qin FG, Yao ZY, et al. Direct ion-beam deposition of carbon-films on silicon in the ion energy-range of 15–500Ev. J Appl Phys 1991;70:5623–7.

[34] Schneider D, Meyer CF, Mai H, Schoneich B, Ziegele H, Scheibe HJ, et al. Nondestructive characterization of mechanical and structural properties of amorphous diamond-like carbon films. Diam Relat Mater 1998;7:973–80.

[35] Druz B, Ostan E, Distefano S, Hayes A, Kanarov V, Polyakov V, et al. Diamond-like carbon films deposited using a broad, uniform ion beam from an RF inductively coupled CH4-plasma source. Diam Relat Mater 1998;7:965–72.

[36] Lu TR, Kuo CT, Yang JR, Chen LC, Chen KH, Chen TM. High purity nano-crystalline carbon nitride films prepared at ambient temperature by ion beam sputtering. Surf Coat Technol 1999;115:116–22.

[37] Paredez P, Marchi MC, da Costa MEHM, Figueroa CA, Kleinke MU, Ribeiro CTM, et al. Carbon nano-structures containing nitrogen and hydrogen prepared by ion beam assisted deposition. J Non-Cryst Solids 2006;352:1303–6.

[38] Li QT, Ni ZC, Gong JL, Zhu DZ, Zhu ZY. Nano-graphite deposits on multi-walled carbon nanotubes induced by low energy ion beam irradiation in a methane and hydrogen mixture. New Carbon Mater 2008;23:235–40.

[39] Yamamoto K, Koga Y, Fujiwara S, Kubota M. New method of carbon nanotube growth by ion beam irradiation. Appl Phys Lett 1996;69:4174–5.
[40] Eletskii AV, Iskandarova IM, Knizhnik AA, Krasikov DN. Graphene: fabrication methods and thermophysical properties. Physics-Uspekhi 2011;54:227–58.
[41] Li X, Cai W, An J, Kim S, Nah J, Yang D, et al. Large-area synthesis of high-quality and uniform graphene films on copper foils. Science 2009;324:1312–4.
[42] Marrs B, Andrews R, Rantell T, Pienkowski D. Augmentation of acrylic bone cement with multiwall carbon nanotubes. J Biomed Mater Res A 2006;77A:269–76.
[43] Nien YH, Huang CL. The mechanical study of acrylic bone cement reinforced with carbon nanotube. Mater Sci Eng B 2010;169:134–7.
[44] Ormsby R, McNally T, Mitchell C, Dunne N. Influence of multiwall carbon nanotube functionality and loading on mechanical properties of PMMA/MWCNT bone cements. J Mater Sci Mater Med 2010;21:2287–92.
[45] Meng DC, Ioannou J, Boccaccini AR. Bioglass(A (R))-based scaffolds with carbon nanotube coating for bone tissue engineering. J Mater Sci Mater Med 2009;20:2139–44.
[46] Boccaccini AR, Chicatun F, Cho J, Bretcanu O, Roether JA, Novak S, et al. Carbon nanotube coatings on bioglass-based tissue engineering scaffolds. Adv Funct Mater 2007;17:2815–22.
[47] Wang ZK, Hu QL, Cai L. Chitosan and multi-walled carbon nanotube composite rods. Chinese J Polym Sci 2010;28:801–6.
[48] Sato Y, Yokoyama A, Kasai T, Hashiguchi S, Ootsubo M, Ogino SI, et al. In vivo rat subcutaneous tissue response of binder-free multi-walled carbon nanotube blocks cross-linked by de-fluorination. Carbon 2008;46:1927–34.
[49] Balani K, Anderson R, Laha T, Andara M, Tercero J, Crumpler E, et al. Plasma-sprayed carbon nanotube reinforced hydroxyapatite coatings and their interaction with human osteoblasts in vitro. Biomaterials 2007;28:618–24.
[50] Meng YH, Tang CY, Tsui CP, Uskokovic PS. Fabrication and characterization of HA-ZrO2-MWCNT ceramic composites. J Compos Mater 2010;44:871–82.
[51] Marrs B, Andrews R, Pienkowski D. Multiwall carbon nanotubes enhance the fatigue performance of physiologically maintained methyl methacrylate-styrene copolymer. Carbon 2007;45:2098–104.
[52] Reis J, Kanagaraj S, Fonseca A, Mathew MT, Capela-Silva F, Potes J, et al. In vitro studies of multiwalled carbon nanotube/ultrahigh molecular weight polyethylene nanocomposites with osteoblast-like MG63 cells. Braz J Med Biol Res 2010;43:476–82.
[53] Shi XF, Hudson JL, Spicer PP, Tour JM, Krishnamoorti R, Mikos AG. Injectable nanocomposites of single-walled carbon nanotubes and biodegradable polymers for bone tissue engineering. Biomacromolecules 2006;7:2237–42.
[54] Shi XF, Sitharaman B, Pham QP, Liang F, Wu K, Billups WE, et al. Fabrication of porous ultra-short single-walled carbon nanotube nanocomposite scaffolds for bone tissue engineering. Biomaterials 2007;28:4078–90.
[55] Kobayashi S, Kawai W. Development of carbon nanofiber reinforced hydroxyapatite with enhanced mechanical properties. Compos Part A 2007;38:114–23.
[56] Lahiri D, Benaduce AP, Rouzaud F, Solomon J, Keshri AK, Kos L, et al. Wear behavior and in vitro cytotoxicity of wear debris generated from hydroxyapatite-carbon nanotube composite coating. J Biomed Mater Res A 2011;96A:1–12.
[57] Elias KL, Price RL, Webster TJ. Enhanced functions of osteoblasts on nanometer diameter carbon fibers. Biomaterials 2002;23:3279–87.

[58] Hirata E, Uo M, Nodasaka Y, Takita H, Ushijima N, Akasaka T, et al. 3D collagen scaffolds coated with multiwalled carbon nanotubes: initial cell attachment to internal surface. J Biomed Mater Res B 2010;93B:544–50.

[59] Zhang HL, Chen ZQ. Fabrication and characterization of electrospun PLGA/MWNTs/hydroxyapatite biocomposite scaffolds for bone tissue engineering. J Bioact Compat Pol 2010;25:241–59.

[60] Singh MK, Gracio J, Leduc P, Goncalves PP, Marques PAAP, Goncalves G, et al. Integrated biomimetic carbon nanotube composites for in vivo systems. Nanoscale 2010;2:2855–63.

[61] Armentano I, Marinucci L, Dottori M, Balloni S, Fortunati E, Pennacchi M, et al. Novel poly(L-lactide) PLLA/SWNTs nanocomposites for biomedical applications: material characterization and biocompatibility evaluation. J Biomat Sci Polym Ed 2011;22:541–56.

[62] Wang W, Watari F, Omori M, Liao S, Zhu YH, Yokoyama A, et al. Mechanical properties and biological behavior of carbon nanotube/polycarbosilane composites for implant materials. J Biomed Mater Res B 2007;82B:223–30.

[63] Abarrategi A, Gutierrez MC, Moreno-Vicente C, Hortiguela MJ, Ramos V, Lopez-Lacomba JL, et al. Multiwall carbon nanotube scaffolds for tissue engineering purposes. Biomaterials 2008;29:94–102.

[64] Armentano I, Alvarez-Perez MA, Carmona-Rodriguez B, Gutierrez-Ospina I, Kenny JM, Arzate H. Analysis of the biomineralization process on SWNT-COOH and F-SWNT films. Mater Sci Eng C 2008;28:1522–9.

[65] Bhattacharya M, Wutticharoenmongkol-Thitiwongsawet P, Hamamoto DT, Lee D, Cui TH, Prasad HS, et al. Bone formation on carbon nanotube composite. J Biomed Mater Res A 2011;96A:75–82.

[66] Mendes RM, Silva GAB, Caliari MV, Silva EE, Ladeira LO, Ferreira AJ. Effects of single wall carbon nanotubes and its functionalization with sodium hyaluronate on bone repair. Life Sci 2010;87:215–22.

[67] Green DE, Longtin JP, Sitharaman B. The effect of nanoparticle-enhanced photoacoustic stimulation on multipotent marrow stromal cells. ACS Nano 2009;3:2065–72.

[68] Tutak W, Chhowalla M, Sesti F. The chemical and physical characteristics of single-walled carbon nanotube film impact on osteoblastic cell response. Nanotechnology 2010;21(31):315102.

[69] Tutak W, Park KH, Vasilov A, Starovoytov V, Fanchini G, Cai SQ, et al. Toxicity induced enhanced extracellular matrix production in osteoblastic cells cultured on single-walled carbon nanotube networks. Nanotechnology 2009;20:255101.

[70] Sitharaman B, Shi XF, Walboomers XF, Liao HB, Cuijpers V, Wilson LJ, et al. In vivo biocompatibility of ultra-short single-walled carbon nanotube/biodegradable polymer nanocomposites for, bone tissue engineering. Bone 2008;43:362–70.

[71] Khang D, Sato M, Price RL, Ribbe AE, Webster TJ. Selective adhesion and mineral deposition by osteoblasts on carbon nanofiber patterns. Int J Nanomed 2006;1:65–72.

[72] Price RL, Waid MC, Haberstroh KM, Webster TJ. Selective bone cell adhesion on formulations containing carbon nanofibers. Biomaterials 2003;24:1877–87.

[73] Khang D, Sato M, Price RL, Ribbe AE, Webster TJ. Selective adhesion and mineral deposition by osteoblasts on carbon nanofiber patterns. Int J Nanomed 2006;1:65–72.

[74] Singh MK, Gracio J, LeDuc P, Goncalves PP, Marques PA, Goncalves G, et al. Integrated biomimetic carbon nanotube composites for in vivo systems. Nanoscale 2010;2:2855–63.

[75] Firkowska I, Godehardt E, Giersig M. Interaction between human osteoblast cells and inorganic two-dimensional scaffolds based on multiwalled carbon nanotubes: a quantitative AFM study. Adv Funct Mater 2008;18:3765–71.
[76] Price RL, Ellison K, Haberstroh KM, Webster TJ. Nanometer surface roughness increases select osteoblast adhesion on carbon nanofiber compacts. J Biomed Mater Res A 2004;70A:129–38.
[77] Sirivisoot S, Yao C, Xiao X, Sheldon BW, Webster TJ. Greater osteoblast functions on multiwalled carbon nanotubes grown from anodized nanotubular titanium for orthopedic applications. Nanotechnology 2007;18:365102.
[78] Sirivisoot S, Yao C, Xiao X, Sheldon BW, Webster TJ. Developing biosensors for monitoring orthopedic tissue growth. Mater Res Soc Symp Proc 2007;950:7.
[79] Sirivisoot S, Webster TJ. Multiwalled carbon nanotubes enhance electrochemical properties of titanium to determine in situ bone formation. Nanotechnology 2008;19(29):295101.
[80] Grausova L, Vacik J, Vorlicek V, Svorcik V, Slepicka P, Bilkova P, et al. Fullerene C-60 films of continuous and micropatterned morphology as substrates for adhesion and growth of bone cells. Diam Relat Mater 2009;18:578–86.
[81] Vandrovcova M, Vacik J, Svorcik V, Slepicka P, Kasalkova N, Vorlicek V, et al. Fullerene C-60 and hybrid C-60/Ti films as substrates for adhesion and growth of bone cells. Phys Status Solidi A-Appl Mater Sci 2008;205:2252–61.
[82] Gonzalez KA, Wilson LJ, Wu W, Nancollas GH. Synthesis and in vitro characterization of a tissue-selective fullerene: vectoring C(60)(OH)(16)AMBP to mineralized bone. Bioorg Med Chem 2002;10:1991–7.
[83] Mirakyan AL, Wilson LJ. Functionalization of C60 with diphosphonate groups: a route to bone-vectored fullerenes. J Chem Soc Perkin Trans 2002;2:1173–6.
[84] Yudoh K, Shishido K, Murayama H, Yano M, Matsubayashi K, Takada H, et al. Water-soluble c60 fullerene prevents degeneration of articular cartilage in Osteoarthritis via down-regulation of chondrocyte catabolic activity and inhibition of cartilage degeneration during disease development. Arthritis Rheum 2007;56:3307–18.
[85] Yudoh K, Karasawa R, Masuko K, Kato T. Water-soluble fullerene (C60) inhibits the development of arthritis in the rat model of arthritis. Int J Nanomed 2009;4:217–25.
[86] Yudoh K, Karasawa R, Masuko K, Kato T. Water-soluble fullerene (C60) inhibits the osteoclast differentiation and bone destruction in arthritis. Int J Nanomed 2009;4:233–9.
[87] Sayes CM, Fortner JD, Guo W, Lyon D, Boyd AM, Ausman KD, et al. The differential cytotoxicity of water-soluble fullerenes. Nano Lett 2004;4:1881–7.
[88] Jia G, Wang HF, Yan L, Wang X, Pei RJ, Yan T, et al. Cytotoxicity of carbon nanomaterials: single-wall nanotube, multi-wall nanotube, and fullerene. Environ Sci Technol 2005;39:1378–83.
[89] Sayes CM, Gobin AM, Ausman KD, Mendez J, West JL, Colvin VL. Nano-C-60 cytotoxicity is due to lipid peroxidation. Biomaterials 2005;26:7587–95.
[90] Zhang Q, Mochalin VN, Neitzel I, Knoke IY, Han J, Klug CA, et al. Fluorescent PLLA-nanodiamond composites for bone tissue engineering. Biomaterials 2011;32:87–94.
[91] Zhang Q, Mochalin VN, Neitzel I, Hazeli K, Niu J, Kontsos A, et al. Mechanical properties and biomineralization of multifunctional nanodiamond-PLLA composites for bone tissue engineering. Biomaterials 2012;33:5067–75.
[92] Pramatarova L, Pecheva E, Stavrev S, Spasov T, Montgomery P, Toth A, et al. Artificial bones through nanodiamonds. J Optoelectron Adv Mater 2007;9:236–9.
[93] Pecheva E, Pramatarova L, Fingarova D, Hikov T, Dineva I, Karagyozova Z, et al. Advanced materials for metal implant coatings. J Optoelectron Adv Mater 2009;11:1323–6.

[94] Mitura S. Nanodiamonds. J Achiev Mater Manuf Eng 2007;24:166–71.
[95] Schrand AM, Hens SAC, Shenderova OA. Nanodiamond particles: properties and perspectives for bioapplications. Crit Rev Solid State Mater Sci 2009;34:18–74.
[96] Huang HJ, Pierstorff E, Osawa E, Ho D. Protein-mediated assembly of nanodiamond hydrogels into a biocompatible and biofunctional multilayer nanofilm. ACS Nano 2008;2:203–12.
[97] Cheng CY, Perevedentseva E, Tu JS, Chung PH, Cheng CL, Liu KK, et al. Direct and in vitro observation of growth hormone receptor molecules in A549 human lung epithelial cells by nanodiamond labeling. Appl Phys Lett 2007;90:163903/1–3.
[98] Shimkunas RA, Robinson E, Lam R, Lu S, Xu XY, Zhang XQ, et al. Nanodiamond-insulin complexes as pH-dependent protein delivery vehicles. Biomaterials 2009;30:5720–8.
[99] Askari SJ. Tribological characteristics of nanocrystalline diamond films grown on titanium. Surf Eng 2009;25:482–6.
[100] Okroj W, Kaminska M, Klimek L, Szymanski W, Walkowiak B. Blood platelets in contact with nanocrystalline diamond surfaces. Diam Relat Mater 2006;15:1535–9.
[101] Fries MD, Vohra YK. Properties of nanocrystalline diamond thin films grown by MPCVD for biomedical implant purposes. Diam Relat Mater 2004;13:1740–3.
[102] Zheng CL, Qi R, Yang WB. MPACVD nanocrystalline diamond for biomedical applications. Key Eng Mat 2005;280–283:1595–8.
[103] Bajaj P, Akin D, Gupta A, Sherman D, Shi B, Auciello O, et al. Ultrananocrystalline diamond film as an optimal cell interface for biomedical applications. Biomed Microdevices 2007;9:787–94.
[104] Grausova L, Kromka A, Bacakova L, Potocky S, Vanecek M, Lisa V. Bone and vascular endothelial cells in cultures on nanocrystalline diamond films. Diam Relat Mater 2008;17:1405–9.
[105] Amaral M, Dias AG, Gomes PS, Lopes MA, Silva RF, Santos JD, et al. Nanocrystalline diamond: in vitro biocompatibility assessment by MG63 and human bone marrow cells cultures. J Biomed Mater Res A 2008;87A:91–9.
[106] Yang L, Sheldon BW, Webster TJ. The impact of diamond nanocrystallinity on osteoblast functions. Biomaterials 2009;30:3458–65.
[107] Yang L, Sheldon BW, Webster TJ. Orthopedic nano diamond coatings: control of surface properties and their impact on osteoblast adhesion and proliferation. J Biomed Mater Res A 2009;91A:548–56.
[108] Kalbacova M, Rezek B, Baresova V, Wolf-Brandstetter C, Kromka A. Nanoscale topography of nanocrystalline diamonds promotes differentiation of osteoblasts. Acta Biomater 2009;5:3076–85.
[109] Broz A, Baresova V, Kromka A, Rezek B, Kalbacova M. Strong influence of hierarchically structured diamond nanotopography on adhesion of human osteoblasts and mesenchymal cells. Phys Status Solidi A-Appl Mater Sci 2009;206:2038–41.
[110] Yang L, Chinthapenta V, Li Q, Stout D, Liang A, Sheldon BW, et al. Understanding osteoblast responses to stiff nanotopographies through experiments and computational simulations. J Biomed Mater Res A 2011;97A:375–82.
[111] Rodrigues AA, Baranauskas V, Ceragioli HJ, Peterlevitz AC, Belangero WD. In vivo preliminary evaluation of bone-microcrystalline and bone-nanocrystalline diamond interfaces. Diam Relat Mater 2010;19:1300–6.
[112] Kloss FR, Gassner R, Preiner J, Ebner A, Larsson K, Hachl O, et al. The role of oxygen termination of nanocrystalline diamond on immobilisation of BMP-2 and subsequent bone formation. Biomaterials 2008;29:2433–42.
[113] Clem WC, Chowdhury S, Catledge SA, Weimer JJ, Shaikh FM, Hennessy KM, et al. Mesenchymal stem cell interaction with ultra-smooth nanostructured diamond for wear-resistant orthopaedic implants. Biomaterials 2008;29:3461–8.

[114] Klauser F, Hermann M, Steinmuller-Nethl D, Eiter O, Pasquarelli A, Bertel E, et al. Direct and protein-mediated cell attachment on differently terminated nanocrystalline diamond. Chem Vapor Depos 2010;16:42–9.
[115] Lechleitner T, Klauser F, Seppi T, Lechner J, Jennings P, Perco P, et al. The surface properties of nanocrystalline diamond and nanoparticulate diamond powder and their suitability as cell growth support surfaces. Biomaterials 2008;29:4275–84.
[116] Pareta R, Yang L, Kothari A, Sirinrath S, Xiao X, Sheldon BW, et al. Tailoring nanocrystalline diamond coated on titanium for osteoblast adhesion. J Biomed Mater Res A 2010;95:129–36.
[117] Yang L, Li YW, Sheldon BW, Webster TJ. Altering surface energy of nanocrystalline diamond to control osteoblast responses. J Mater Chem 2012;22:205–14.
[118] Kromka A, Grausova L, Bacakova L, Vacik J, Rezek B, Vanecek M, et al. Semiconducting to metallic-like boron doping of nanocrystalline diamond films and its effect on osteoblastic cells. Diam Relat Mater 2010;19:190–5.
[119] Jakubowski W, Bartosz G, Niedzielski P, Szymanski W, Walkowiak B. Nanocrystalline diamond surface is resistant to bacterial colonization. Diam Relat Mater 2004;13:1761–3.
[120] Hauert R. A review of modified DLC coatings for biological applications. Diam Relat Mater 2003;12:583–9.
[121] Polcar T, Vitu T, Cvrcek L, Novak R, Vyskocil J, Cavaleiro A. Tribological behaviour of nanostructured Ti-C:H coatings for biomedical applications. Solid State Sci 2009;11:1757–61.
[122] Yasumaru N, Miyazaki K, Kiuchi J. Control of tribological properties of diamond-like carbon films with femtosecond-laser-induced nanostructuring. Appl Surf Sci 2008;254:2364–8.
[123] Ma WJ, Ruys AJ, Zreiqat H, Mason R, Ringer SP, Liu ZW, et al. The biocompatibility of diamond-like carbon nano films. In: Nanoscience and Nanotechnology, 2006 ICONN '06 International Conference on 2006; 2006.
[124] Dorner-Reisel A, Schürer C, Nischan C, Klemm V, Irmer G, Müller E. Calcium-oxygen modified amorphous and nanocrystalline carbon layers as biomaterials. Biomed Tech (Berl) 2002;47:393–6.
[125] Fan H, Wang L, Zhao K, Li N, Shi Z, Ge Z, et al. Fabrication, mechanical properties, and biocompatibility of graphene-reinforced chitosan composites. Biomacromolecules 2010;11:2345–51.
[126] Czarnecki JS, Lafdi K, Tsonis PA. A novel approach to control growth, orientation, and shape of human osteoblasts. Tissue Eng Part A 2008;14:255–65.

Self-assembled nanostructures for bone tissue engineering

6

Bone possesses high regenerative capacity, particularly for younger people, and this capacity means that bone can heal fracture itself in some circumstances such as the fracture and defect smaller than critical sizes. This regenerative capacity has been utilized to treat large-size or complicated discontinuity problems in bone with the assistance of autogenic bone, biomaterials, or cell therapy, which has evolved into a new biomedical and bioengineering field called bone tissue engineering. In contrast to conventional biomaterial approaches for treating orthopedic problems by implanting prosthetic replacements or fixation parts, the tissue engineering approach is based on the in-depth understanding of tissue formation and regeneration, and aims to induce new functional tissue by exploring the potentials of scaffold materials, cells, or biomolecules (e.g., growth factors, proteins) in a synergetic way [1]. Because of this nature, bone tissue engineering requires spatiotemporally balanced and coordinated combinations between different components at least including material scaffold, cells, and culturing proteins or growth factors (illustrated in Figure 6.1). Despite of the complexity raised from the interaction of multiple components and problems of efficacy and insufficient human tests, tissue engineering has exhibited a great success in laboratory and preclinical trials. With the assistance of nanotechnology, the efficacy and potential of tissue engineering for treating orthopedic diseases have been substantially improved and the applicable scope of bone tissue engineering has been greatly enlarged.

In previous chapters, several nanomaterials for bone tissue engineering and their potentials in research and clinics have been mentioned. Here, a new area of bionanotechnology, tissue engineering enabled by self-assembled nanostructures, is discussed and introduced with details. Self-assembled nanostructures represent a unique group of materials or matters beyond traditional forms of material (i.e., metals, polymers, ceramics), possessing the intrinsic ability to "automatically" build hierarchical or ordered architectures in a way that biological system does. This biomimetic nature makes self-assemblies excellent scaffolds for tissue engineering applications and cell therapy, which will be elucidated in this chapter. Also, principles for self-assembly and fabrication methods of tissue engineering scaffolds will be covered.

6.1 Fabrication of self-assembled nanostructures

6.1.1 Principles for self-assembly

Since many biological systems use self-assembly to construct various naturally occurring molecules and structures, nanotechnologists imitate this strategy to fabricate

Nanotechnology-Enhanced Orthopedic Materials. http://dx.doi.org/10.1016/B978-0-85709-844-3.00006-9

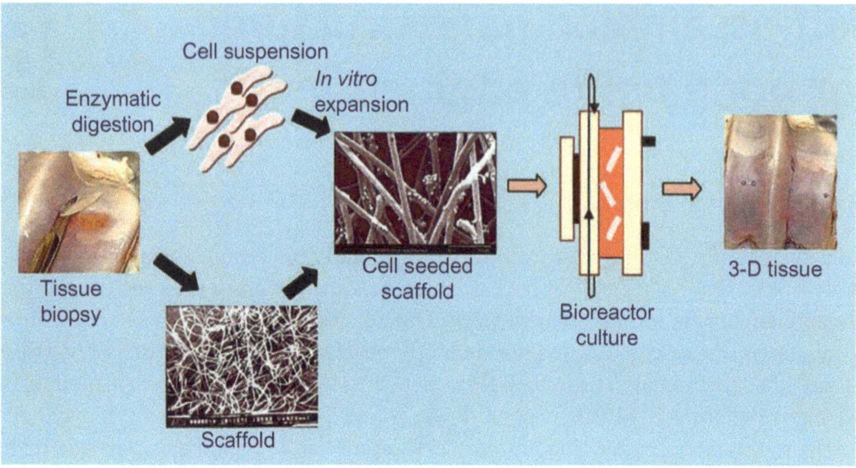

Figure 6.1 An illustration of the typical process of bone tissue engineering strategy for the repair of cartilage defects.
Adapted from Ref. [2] with permission.

nanostructures or nanomaterials like supramolecular assemblies and nanofibrous structures. Self-assembly is defined as reversible processes of disordered molecules or tiny building blocks form patterned or ordered structures, mostly occurring spontaneously and thermodynamically driven by multiple noncovalent interactions. These interactions include hydrogen bonding, electrostatic association and van der Waals forces, and are often regulated by the hydrophobic/hydrophilic or capillary effects [3]. Self-assembly is usually classified to static and dynamic types, between which static self-assembly is better known. The difference between the two types of self-assembly is whether there is energy dissipation during the interaction of components: static self-assembly has no energy dissipation while the dynamic self-assembly does [4]. Figure 6.2 illustrates a number of typical static self-assemblies.

There are number of common features shared by self-assemblies [4]. First, self-assembly reflects information coded in individual components. The information can be shape, surface properties, charge, polarity, mass, and so on, and such characteristic information determines the interactions between components. Second, the components are mobile with respect to each other, driven by the attraction or repulsion as a result of the interactions between components. Third, the self-assembly process usually occurs in fluid phases or on smooth surface which assist the movement of components. Finally, self-assembling components should be able to reach equilibrium between aggregated and nonaggregated states, or adjust their relative position once in an aggregate.

These common features suggest that self-assembling process is actually a complex yet organized process mediated by many factors, which in turn should be carefully considered in the manufacture of self-assembled architectures. Specifically, these crucial factors include (but are not limited to) molecular (or component) mobility, proximity to equilibrium, and anisotropy [5]. Mobility refers to the ability of molecules or components to change position and orientation, which is usually

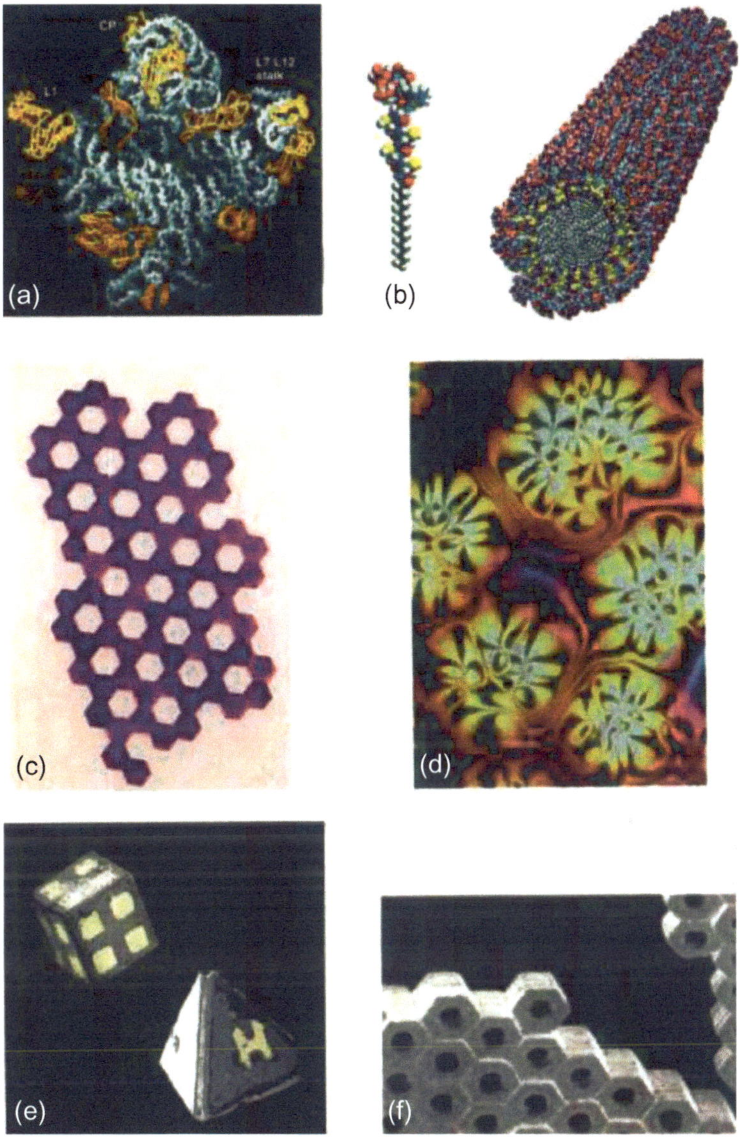

Figure 6.2 Examples of static self-assembly. (a) Crystal structure of a ribosome. (b) Self-assembled peptide-amphiphile nanofibers. (c) An array of millimeter-sized polymeric plates assembled at a water/perfluorodecalin interface by capillary interactions. (d) Thin film of a nematic liquid crystal on an isotropic substrate. (e) Micrometer-sized metallic polyhedral folded from planar substrates. (f) A 3-D aggregate of micrometer plates assembled by capillary forces.
Reprinted with permission from Ref. [4].

sensitive to an aqueous environment and temperature. High mobility and short transportation paths for molecules can accelerate assembly rate and, subsequently, growth rates of nanostructures. Proximity to equilibrium also plays an important role in regulating assembly rates, perfection of structures, and stability or purity of resultant materials. In general, near-equilibrium conditions cause slow growth or low assembly rate but high perfection and reproducibility in the self-assembled structures, as well as low susceptibility to structural or compositional changes and property degradation. Anisotropy is a determinant for the geometry and structure of self-assembled materials and it can be imposed both externally and internally or raised from a favorable environment. The external stimuli to induce anisotropy include directional flow and reagent gradient, while the internal stimuli usually refer to template or selective nucleation sites. An example of the favorable environment for generating anisotropy is the liquid phase that causes uneven distribution of charge at molecular surfaces, driving the building blocks aggregate and grow into anisotropic supramolecular structures.

6.1.2 Fabrication strategy of self-assembled nanostructures

As mentioned above, information coded in the components of self-assembly determines the way of the components responding to or interacting with their surroundings. In general, the information can be geometrical (dimension, shape, size, etc.), physiochemical (surface properties, charge, polarity, mass, etc.), and structural (anisotropy, heterogeneity, etc.). Therefore, designing a self-assembling strategy of novel nanostructures largely relies on the understanding of such information in building blocks. Here, a number of well-established strategies for fabricating various types of self-assembled nanostructures are introduced.

6.1.2.1 Peptide-based nanostructures

Due to many attractive properties, peptides are the most commonly used building blocks and the most comprehensively understood for constructing self-assembled biomaterials. Peptides are reasonably tractable and scalable compared to other moiety for self-assembly and, in terms of biological functionality and flexibility, peptides are highly versatile and facile [3]. In most peptide-based self-assemblies, β-sheet and α-helix are frequently used for assembling nanostructures and, thus, become important basic structures for fabricating self-assembled nanostructures.

It is well known that peptides containing alternating hydrophobic and hydrophilic amino acids often assemble into β-sheet. This understanding has been used to fabricate β-sheet tapes or ribbons, which subsequently pack into sheets or form fibers at nanoscale. Figure 6.3a and b illustrate the sheet, ribbon, and fibers nanostructures formed by β-sheet through complementary ionic interaction and interaction of alternative hydrophilic/hydrophobic residues, respectively. Another strategy using β-sheet to construct self-assembly nanostructures is based on triblock peptides with ABA motif-structures, where A represents a positively charged segment (e.g., lysine residues) and B is an amphiphilic part. As illustrated in Figure 6.3c, the ABA triblock

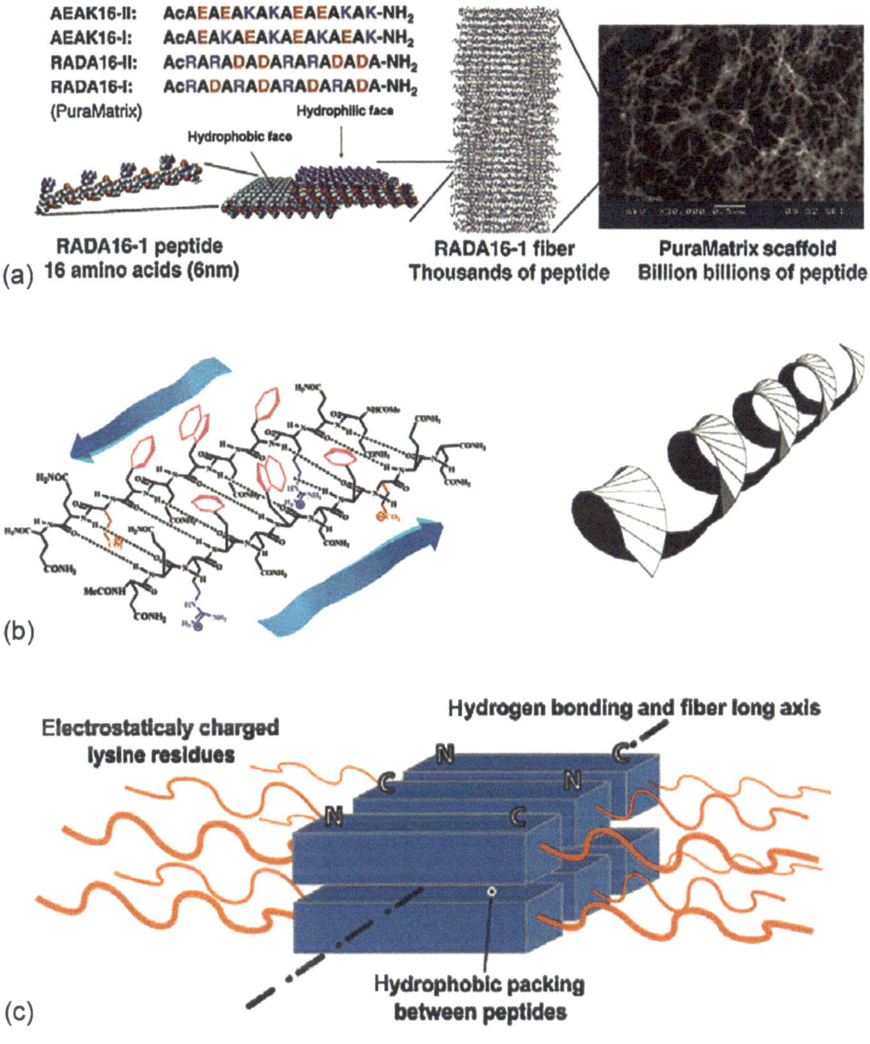

Figure 6.3 Self-assembled nanostructures based on β-sheet: (a) peptides packing into sheets and fibers based on hydrophobic interactions on one face of the molecule, and complementary ionic interaction on the other; (b) peptides with alternating hydrophilic and hydrophobic residues assemble into β-sheet structures (left) that form twisted ribbons (right) and bundle into larger fibers; and (c) self-assembly based on amphiphilic triblock peptides, where the central hydrophobic block forces self-assembly via hydrophobic interactions between molecules and hydrogen bonding along the fiber axis.
Modified with permission from Ref. [3].

peptides form dimers due to hydrophobic interaction between A segments and the dimers assemble perpendicularly to the peptide backbone axis through hydrogen bonds in β-sheet.

Similar to β-sheet, α-helix has also been extensively studied for self-assembly and used for fabricating self-assembled nanostructures. In a general strategy, peptides biomimetically mimicking the helices in natural proteins are synthesized first, and α-helix formed by these peptides can further self-assemble into higher-order structures with, for example, a coiled coil conformation [6]. Due to space limitations, readers can refer to other sources for more information about α-helix (references [3,6], etc.).

Peptide-based self-assembly is capable of synthesizing nanomaterials with various microstructures from tubes to spheres. Dipeptides, probably the simplest building blocks with a versatility in controlling microstructures, can form tubular, spherical, fibrous, or bundle, and array structures by modifying motifs on the dipeptides and/or altering synthesis conditions (Figure 6.4a). For example, the Alzheimer's β-amyloid peptide with diphenylalanine motif forms multiwalled nanotubes when the peptide is

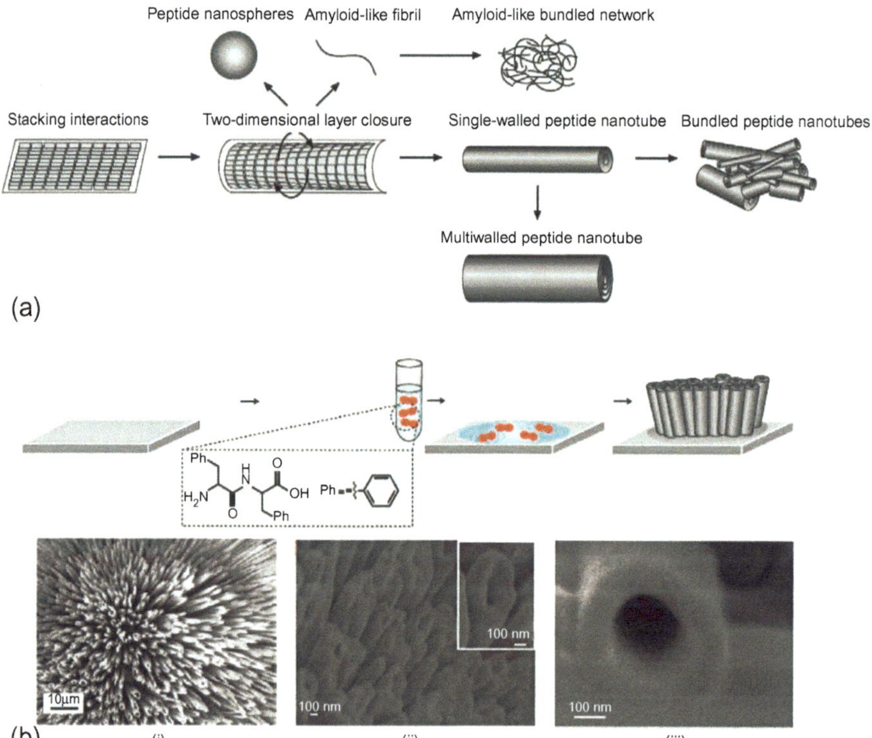

Figure 6.4 (a) Schematic of the formation of tubular (single or multiwalled), spherical, or fibrillar structures via dipeptide self-assembly. (b) Proposed model for the formation of aligned peptide nanotube arrays. Scanning electron micrograph of (i) the vertically aligned peptide nanotubes, (ii) the nanotube arrays, and (iii) an individual nanotube obtained. Reprint with permission from Ref. [7].

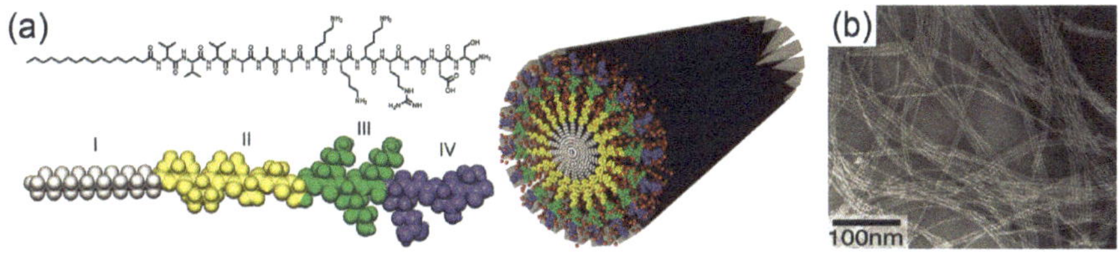

Figure 6.5 (a) Chemical structure of a representative peptide amphiphile (PA) and (b) TEM image of its assembly into nanofibers.
Reprinted from Ref. [9] with permission.

dissolved at 100 mg/ml in 1,1,1,3,3,3-hexafluoro-2-propanol (HFIP) and then diluted down with water at a final concentration of ≤2 mg/ml [7]. Alternatively, spherical nanoparticles form when a similar but more rigid diphenylalanine peptide is used. In addition, when a thiol is functionalized on the diphenylalanine peptide, nanospheres rather than nanotubes can form.

Directional arrays of dipeptide nanotubes (or nanoforests) can be fabricated by simply dissolving the dipeptide in HFIP onto siliconized glass followed by evaporation, as illustrated in Figure 6.4b. This directional growth of self-assembled nanotubes was attributed to the additive assembling of peptides at the growing end of tubes and formation of multiple nucleation sites facilitated by the supersaturation state after solvent evaporation. In addition, horizontally aligned peptide nanotubes were fabricated by using noncovalent coating of magnetite nanoparticles and an external magnetic field [8]. These results indicate that dipeptide nanotubes are capable of forming 2-D self-assembled structures.

Modification of natural peptides by introducing extra functionality or chemistry has been demonstrated to be an efficacious way to alter the nanostructure formation and also a acomplement to peptides intrinsic tendency of self-assembly. One of the most important examples is peptide amphiphile (PA) that is a peptide sequence with an appended alkyl tail and can readily self-assemble into high-aspect ratio fibers. Figure 6.5 shows the molecular structure of a representative PA and the morphology of its assembly. This PA contains four different regions (Figure 6.5a): region I is alkyl (usually C_{16}) tail; region II is a β-sheet forming segment to promote H-bonding along the fiber axis; region III contains charged amino acids; and region IV is a peptide epitope [9]. This molecular structure enables the PAs to self-assemble into high aspect ratio cylindrical nanofibers in aqueous solution, and the nanofibers are about 6 nm in diameter and often several microns in length (Figure 6.5b). It is also important to emphasize that many types of PA are also capable of biomineralization (see Chapter 4).

6.1.2.2 Protein- and polypeptide-based nanostructures

Protein and large polypeptides usually possess more functionality facilitating self-assembly than peptides. A number of natural proteins (including collagen, elastin, silk protein, etc.) that spontaneously self-assemble into high-order structures have been widely used in the laboratory to fabricate self-assembled nanostructures for biomedical applications. Collagen-based nanoassembly has been introduced in the section about biomineralization in Chapter 4. Another well-known natural protein with a long application history, silk protein or specifically silk fibroin, has also been developed for fabricating self-assembled nanostructures for biomedical uses. The silk fibroin consists of a light chain (M_w ~ 26 kDa) and a heavy chain (M_w ~ 390 kDa) linked by a disulfide bond, which makes it a block copolymer rich in hydrophobic β-sheet-forming blocks linked by small hydrophilic linker segments [10]. The crystalline regions of silk fibroin are primarily composed of glycine-linked peptides, including glycine-alanine, -serine, -threonine, or -valine. This structure leads to a hydrophobic protein that can self-assemble to form strong and resilient materials. The high mechanical strength and toughness results from the dominance of the β-sheet regimes within the fibroin. The toughness of silk fibers is possibly greater than the best synthetic materials, while in terms of strength, silk fibers are superior to commonly used polymeric degradable biomaterials such as collagen and poly(L-lactic acid) (PLA), reaching 740 MPa (ultimate tensile strength). In contrast, collagen and PLA have ultimate tensile strengths of 0.9–7.4 MPa and 28–50 MPa, respectively [11]. Figure 6.6 illustrates a fabrication method for silk hydrogel. In the process, starting silk fibroin gel is extracted from silk cocoons by a method described elsewhere [10]. The silk fibroin gel is then vortexed and sonicated and followed by treating with direct electrical current or lowering the pH.

Besides the naturally occurring proteins, coiled-coil protein, amphiphilic block-copolypeptide, and two-component protein are also used for constructing self-assemblies with interest to biomaterial and biomedical engineering fields [3]. Figure 6.7 depicts the self-assemblies built from these proteins or polypeptides.

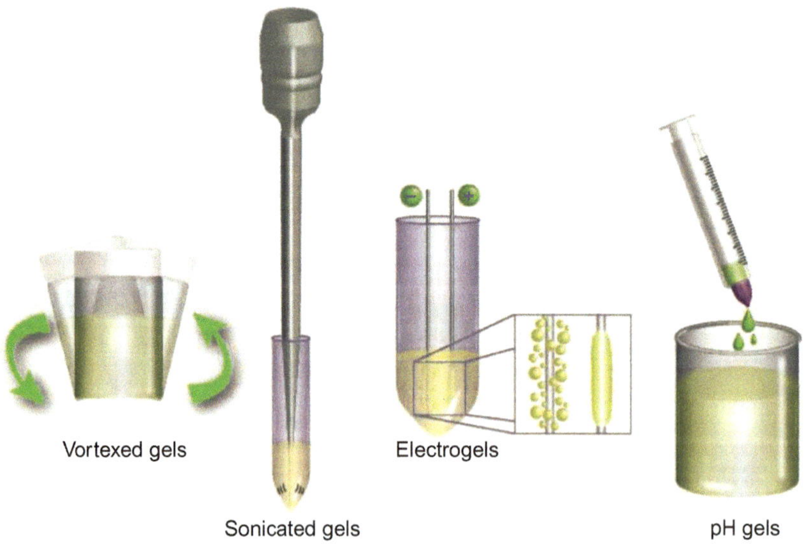

Figure 6.6 Method of preparing self-assembled silk hydrogel.
Reprinted with permission from Ref. [10].

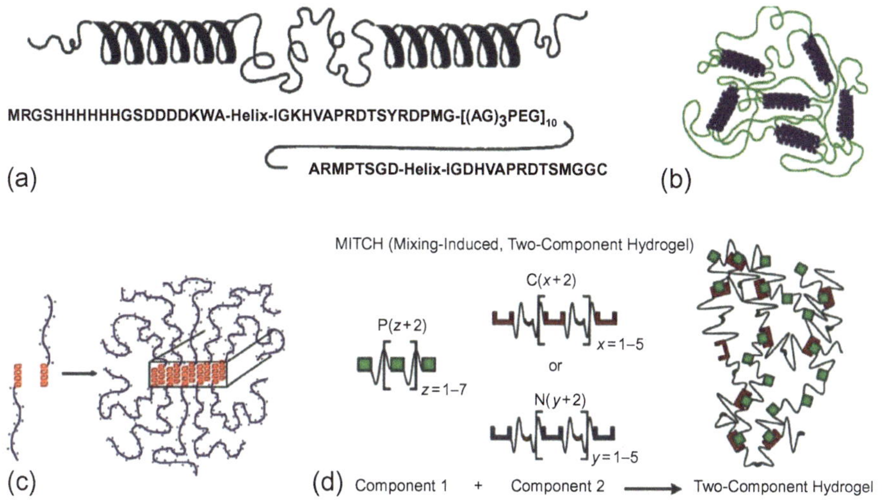

Figure 6.7 Protein-based self-assembled nanomaterials and fabrication strategies: (a) structure of coiled-coil polypeptide, (b) association of the end helices into coiled-coils results in hydrogel formation, (c) amphiphilic block copolymers assemble into extended tapered end fibers, forming hydrogels, and (d) association of two-component proteins to form a hydrogel. Modified from Ref. [3] with permission.

Figure 6.7a shows triblock polypeptides with leucine zipper motifs separated by an unstructured amino acid sequence, which associate into coiled-coils and form hydrogel (Figure 6.7b). In the employment of amphiphilic block copolymers for synthesizing self-assembled nanostructures, the hydrophobic segments in the copolymers form α-helices or β-sheets that, when dissolved in water, do not form micelles but self-assemble via hydrophobic packing into extended tapes or fibers that form hydrogels (Figure 6.7c). In the method of using two-component protein that is first developed by Heilshorn et al., two naturally occurring protein motifs, WW domain and proline-rich peptide, which can associate via noncovalent interactions, are able to form a highly cross-linked network upon mixing (Figure 6.7d). Moreover, the mechanical strength of the resulting self-assembled hydrogel is adjustable by the number of repeats and interaction strength of the two components in the protein [12].

6.1.2.3 DNA-based nanostructures

Although there are only a few existing examples of DNA-based biomaterials due to high costs, the method of using DNA to construct self-assembled nanostructures is promising and has lots of potential. This is mainly attributed to the programmable nature of DNA that any two DNA strands of appropriate sequence can assemble via base pairing, which allows for great flexibility in constructing desirable nanostructures. Here, the example of helical rosette nanotubes (HRNs) is discussed. HRN is a type of soft organic nanomaterials obtained through the self-assembly of low molecular weight synthetic DNA molecules. HRNs with proper functionality have been

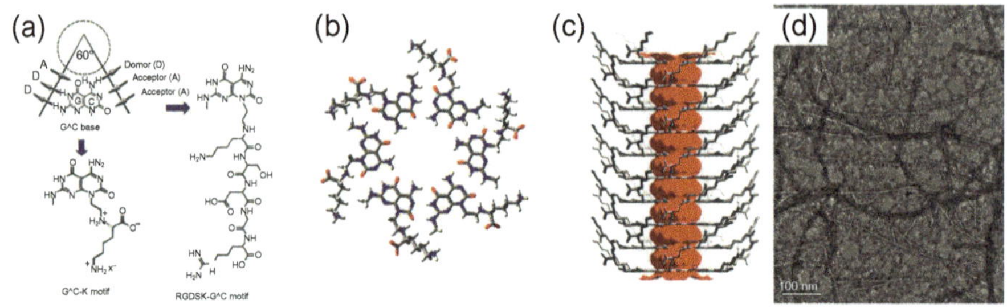

Figure 6.8 Schematic of the self-assembly of DNA-based HRN (a–c) and (d) transmission electron microscopy (TEM) image of HRN.
Modified and redrawn from Ref. [14].

developed to mimic the natural nanostructures of collagen and other components in bone (Figure 6.8). Building blocks of HRN are G^C base, G^C with L-lysine side chain (G^C-K), and synthesized RGD side chain conjugated G^C (RGD-K-G^C) (as illustrated in Figure 6.8a), and six of such building blocks assemble into rosette-like supermacrocycle (Figure 6.8b). In the aqueous environment, these rosettes further stack up via hydrophobic, electrostatic, and base pairing interactions (Figure 6.8c) and eventually form stable helical nanotubes (Figure 6.8d). A study also demonstrated that HRN can be readily coated on titanium or incorporated in hydroxyapatite (HA) [13].

6.2 Applications of self-assembled nanostructures for bone tissue engineering

In the past decade, there are a number of promising self-assembled nanostructures with attractive properties and great potential for bone tissue engineering applications. These nanostructures of interest are in the forms of hydrogels or scaffolds consisting of nanotubular or nanofibrous materials fabricated by the aforementioned methods. Here, some typical self-assembled nanostructures for bone tissue engineering are introduced.

6.2.1 *Helical rosette nanotube* (HRN)

As mentioned before, HRN is DNA-based, helical nanofibers in the form of hydrogel that possesses attractive features for orthopedic tissue engineering. These features include (1) HRN has rich amino acid side chains with well-controlled spatial nanoscale distributions that impart more functionality and versatility to meet a wide range of orthopedic needs; (2) HRN hydrogel is able to solidify when heated or added directly into a serum-free medium, which allows them to serve as injectable tissue engineering materials; and (3) HRN has demonstrated potential to enhance osteoblast adhesion and serve as an excellent calcification template [13,14]. Based on these unique properties, HRN has been studied or incorporated with other materials or molecules to create a better implantable material for enhancing bone regeneration.

For example, self-assembled HRN conjugated with RGD peptides (peptides responsible for cell adhesion) as well as lysine was incorporated into solidified hydrogels to improve the osteoblast functions and cytocompatibility of hydrogels [14]. Specifically, HRNs, RGD-K, and/or L-lysine peptides (HRN-RGD-K or HRN-K) were codissolved in deionized water and 2-hydroxyethyl methacrylate (HEMA) monomers were added in the solutions with free radical initiators (2, 2'-azobisisobutyronitrile). This slurry was kept at 78 °C for less than 1 hour for polymerization of HEMA. Then the polymerized HEMA (pHEMA) hydrogels embedded with functionalized HRNs was cooled down, cut into disk-shape substrates, and compared osteoblast adhesion to uncoated, collagen, and poly L-lysine-coated pHEMA hydrogels. The results in Figure 6.9a and b provide evidence of enhanced (up to twofold) osteoblast adhesion on 10% HRN-RGD-K-coated hydrogels than on uncoated hydrogels and 73% more than on collagen-coated hydrogels of similar concentration. Moreover, osteoblast adhesion was greater on 0.01 mg/ml HRN-K-coated hydrogels compared to the same concentration of poly L-lysine-coated hydrogels (Figure 6.9b). These enhancements of cell adhesion were correlated to the changes in surface property when using HRN conjugated with peptides. Figure 6.9c showed the nanoscale features of assembled HRNs and it is clear that embedment of these self-assembled nanostructures can increase roughness. Besides, coupling cell adhesive peptides like RGD apparently changed surface chemistry. These two effects together contributed to the change of surface energy. However, in this case, these factors revealed their complexity of affecting surface energy and protein adsorptions. Specifically, increased fibronectin adsorption was observed on RGD and lysine conjugated HRNs compared to blank hydrogel controls, but 10% HRN-RGD-K-coated hydrogels had comparable fibronectin adsorption compared to hydrogels conjugated with non-RGD modified HRN (HRN-K). One explanation for this result was the countereffects between high surface roughness for 10% HRN-RGD-K-coated hydrogels (which decreases surface energy, especially when roughness is in the submicron regime) and immobilized RGD peptides (which may increase surface energy). In summary, the results of research in self-assembled nanomaterial suggest an alternative way to achieve favorable surface chemistry and nanobiomimetic organic structures to improve bone cell responses.

6.2.2 Hybrid nanostructures based on PA self-assembly

Biomineralization property and potential application of self-assembled PA in promoting cellular responses and calcium phosphate deposition have been briefly introduced in Chapter 4. Due to these promising features, PA represents a group of bioactive self-assembling peptides that are of great potential to bone tissue engineering. Along this line, PA self-assembled nanostructures are often integrated with other molecules or materials of specific functionality to create hybrid scaffolds for the encapsulation of cells and the delivery of osteoinductive or osteoconductive molecules.

For example, PA self-assembly is hybridized with fibronectin epitope RGDS (cell adhesion molecule) and HA crystals that nucleate around the phosphoserine residues of PA under appropriate physiological conditions (referring to the biomineralization section in Chapter 4), forming 3-D biomimetic systems as a matrix to promote bone regeneration [15]. Efficacy of this tissue engineering scaffold was tested in an rat

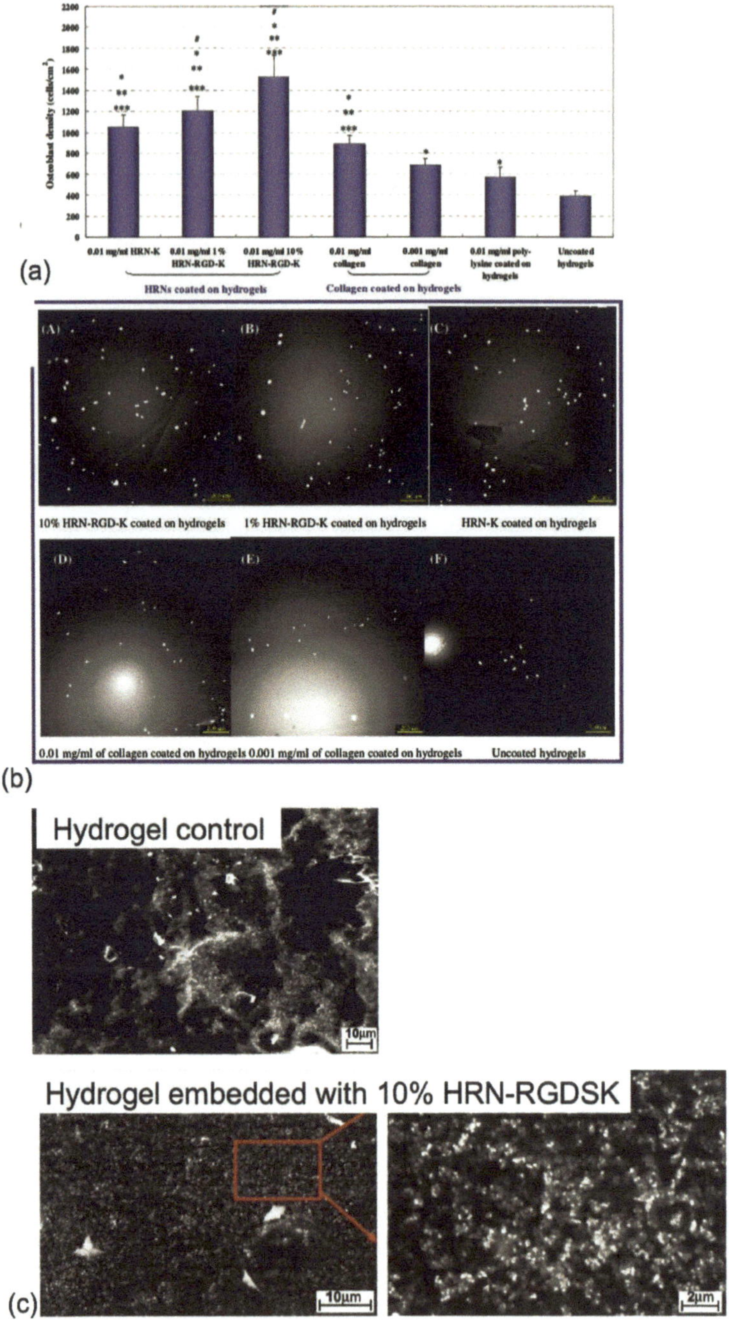

Figure 6.9 (a) Enhanced osteoblast adhesion on HRN-RGD-K-coated hydrogels compared to collagen-coated hydrogels. *$p < 0.05$ compared to uncoated hydrogels, **$p < 0.05$ compared to 0.001 mg/ml collagen-coated hydrogels, ***$p < 0.01$ compared to 0.01 mg/ml poly L-lysine-coated hydrogels, #$p < 0.05$ compared to 0.01 mg/ml collagen-coated hydrogels. (b) Fluorescence microscopy images of stained nuclei of osteoblasts on different hydrogels. (c) Surface morphologies of 0.01 mg/ml 10% HRN-RGD-K hydrogels and hydrogel control without modification (SEM images).
Modified with permission from Ref. [14].

femoral critical-size defect model. The results of bone regeneration *in vivo* are shown in Figure 6.10a and it is clear that animals treated with the self-assembled PA with phosphoserine residues (i.e., S(P)-PA) exhibited statistically higher volumes of bone relative to the untreated group. Furthermore, animals treated with RGDS + S(P)-PA also exhibited significantly higher bone volume than those treated with nonbioactive PA (S-PA) or RGDS + Filler-PA, exhibiting similar values of regenerated bone as those treated with allogenic demineralized bone matrix (DBM). Figure 6.10b shows imaging evidence from microCT, which clearly demonstrates greater new bone formation around RGDS + S(P)-PA compared to untreated bone defects.

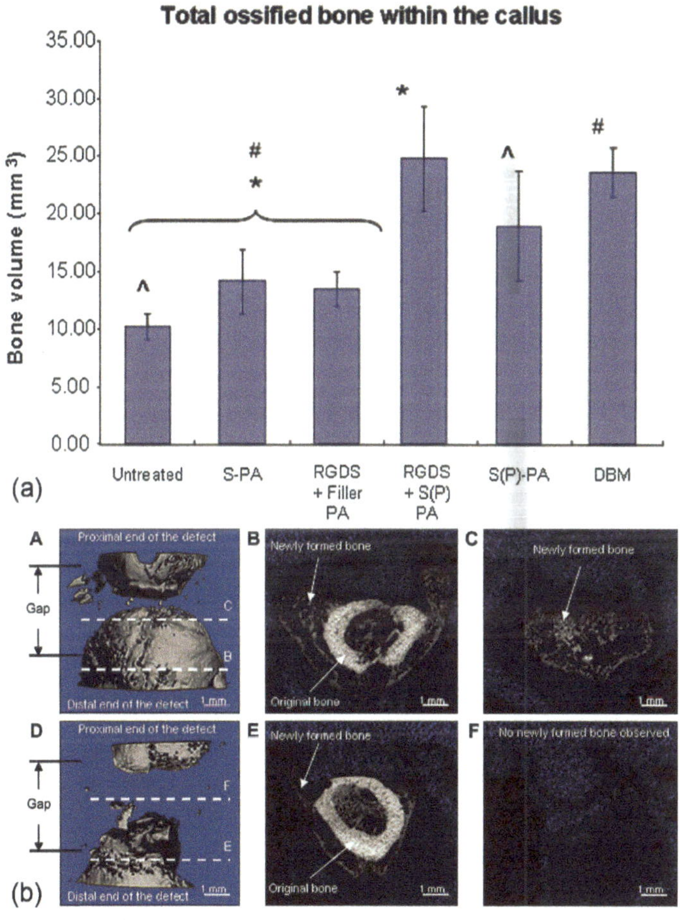

Figure 6.10 (a) Quantification of the *in vivo* bone regeneration efficacy of S(P)-PAs. Microcomputed tomography quantification of bone regeneration. Total volume of bone tissue within the callus quantified by microcomputed tomography (microCT) scans at 4 weeks postoperative. (b) MicroCT imaging of newly formed bone. Representative microCT images depicting the difference in bone formation observed in animals implanted with RGDS+S(P)-PA within the defect (A–C) versus those left untreated (D–F). Modified from Ref. [15] with permission.

Since heparin is a natural biopolymer that can interact with specific binding domains on several growth factors, heparin-binding PA (HBPA) has been developed to bear and deliver a variety of growth factors for tissue regeneration purposes [16]. For example, nanostructured heparin-PA was devised to promote angiogenesis by releasing vascular endothelial growth factor (VEGF) and basic fibroblast growth factor (bFGF). Given that angiogenesis is important for bone growth and remodeling, such hybrid nanoscaffolds are of particular interest to bone engineers. Figure 6.11 shows the evaluation results of angiogenesis ability of four different HBPA membranes (with or without 0.5 wt% heparin and with or without added bFGF and VEGF) from a chicken chorioallantoic membrane (CAM) assay. Within 1 day, perimembrane vessel density or vessel morphology were similar in the nonheparin-containing membrane without (Figure 6.11a, B) or with (Figure 6.11a, D) growth factors. The same results were observed for membranes prepared with 0.5 wt% heparin without growth factors compared to the nonheparin-containing membranes. However, the 0.5 wt% heparin membrane with VEGF and bFGF induced a dramatic increase in vessel number, diameter, and morphology in the perimembrane vasculature after 1 day, showing "corkscrew" morphologies of vessels that are a hallmark of VEGF signaling (Figure 6.11a, H). Figure 6.11b is the quantitative result of vessel density on four different membranes. The 0.5 wt% heparin membrane with growth factors resulted in a significant increase in vessel density compared to all other groups after only 1 day.

This heparin-hybridized PA system reveals the ability to promote rapid angiogenesis by utilizing heparin-binding growth factors involved in signaling for angiogenesis. In a different strategy, transforming growth factor-β1 (TGF-β1) was directly bound to a PA nanofiber using a binding sequence derived from phage display [18]. This hybrid PA gel bearing TGF-β1 was shown to promote cartilage regeneration in a rabbit model.

Due to its attractive osteoconductive or osteoinductive properties, PA self-assembly with high bioactivity or extra functionality becomes a promising modifier to currently used bone tissue engineering scaffolds. For example, self-assembled PA nanofiber is added to collagen-poly(glycolic acid) (PGA) sponge to achieve enhanced bone regeneration [19]. Such a hybrid scaffold was tested with the combination of bFGF for its ability to accelerate bone regeneration. The results of *in vivo* assessment of ectopic bone formation are shown in Figure 6.12, where significantly improved bone regeneration was observed in the collagen-PGA scaffold hybridized with PA self-assembled nanofibers compared to collagen-PGA alone or bFGF control. These results strongly suggest that the controlled release of bFGF and other growth factors (such as bone morphogenetic proteins) from collagen-PGA/self-assembled PA nanofiber hybrid scaffold can play an important role in creating an appropriate environment for inducing bone regeneration. It also demonstrated the capability of self-assembled nanostructures in modifying the existing tissue engineering scaffolds or materials for the delivery of drugs and the enhancement of bone regeneration.

In the last example, PA nanofibrous matrix was filled to Ti-6Al-4V foam to obtain a PA-Ti hybrid implant in order to overcome the problem of weak osteoinductive and osteoconductive abilities of Ti alloy [20]. This approach combined the

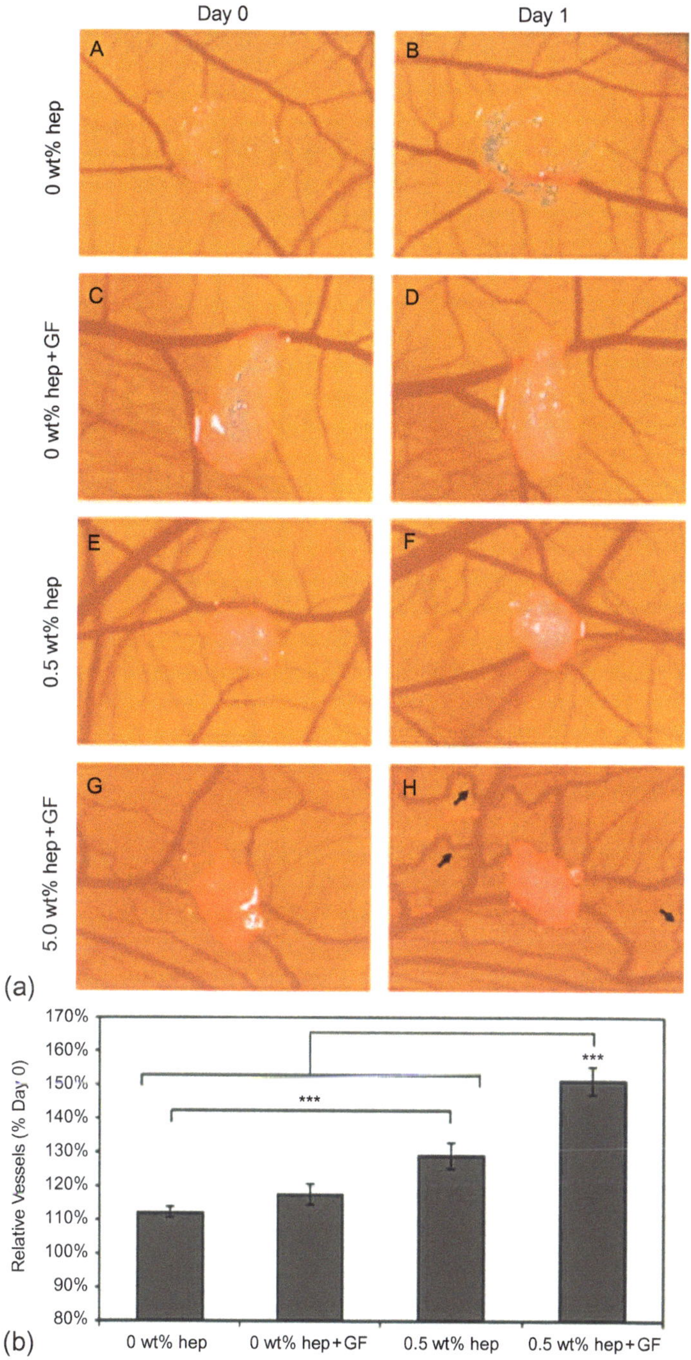

Figure 6.11 (a) Digital images of membranes self-assembled from 2 wt% HBPA and 1 wt% hyaluronic acid with (A, B) 0 wt% heparin, (C, D) 0 wt% heparin with VEGF and bFGF, (E, F) 0.5 wt% heparin, and (G, H) 0.5 wt% heparin with VEGF and bFGF on the chorioallantoic membrane on days 0 and 1. Arrows in (H) indicate vessel morphology indicative of VEGF signaling. (b) Vessel density relative to day 0 induced by hyaluronic acid/HBPA membranes with 0 wt% or 0.5 wt% heparin with or without VEGF and bFGF (***$p < 0.001$). Modified from Ref. [17] with permission.

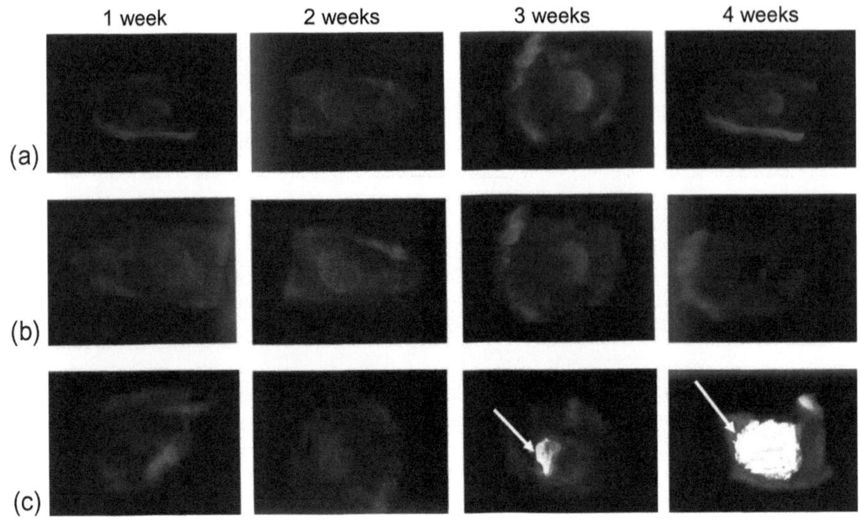

Figure 6.12 Soft X-ray photographs of ectopically formed bone 1, 2, 3, and 4 weeks after implantation of collagen sponges reinforced with the incorporation of PGA fiber, incorporated with PA (a), basic fibroblast growth factor (bFGF) (b), and self-assembled PA nanofibers (hybrid scaffold) (c). Arrows indicate the newly formed bone. The concentration of bFGF used was 50 μg.
Reprinted with permission from Ref. [19].

advantage of tissue engineering strategy with the conventional bone prosthesis, leading to a solution to repair the large defects over critical sizes in load-bearing bones. Specifically, the PA nanofiber matrix self-assembled in situ within the Ti-6Al-4V foam, allowing for the encapsulation of preosteoblastic cells as well as *in vivo* nucleation and subsequent deposition of calcium phosphate (due to phosphoserine residue in the PA). These treatments can significantly enhance the osteoinductive and osteoconductive properties of the metal foam, as demonstrated in Figure 6.13. When tested in a rat femoral model, new and highly mineralized bone with the same structure as cortical bone was observed adjacent to as well as inside the modified metal foam (Figure 6.13e and f).

6.3 Summary and future directions

To date, self-assembled nanostructures have been mainly developed for tissue engineering scaffolds in the form of hydrogel with the extended ability to deliver functional molecules of interest. However, the unique nature and properties of self-assembly and its fabrication process impart these kinds of materials many advantages for promoting bone regeneration and tissue repair. In this chapter, the development of typical self-assembling peptides or DNA-based hydrogels and scaffolds are highlighted and their capability to promote cell growth and tissue formation is discussed. With regards to the ideal properties of tissue engineering scaffold, it has already been

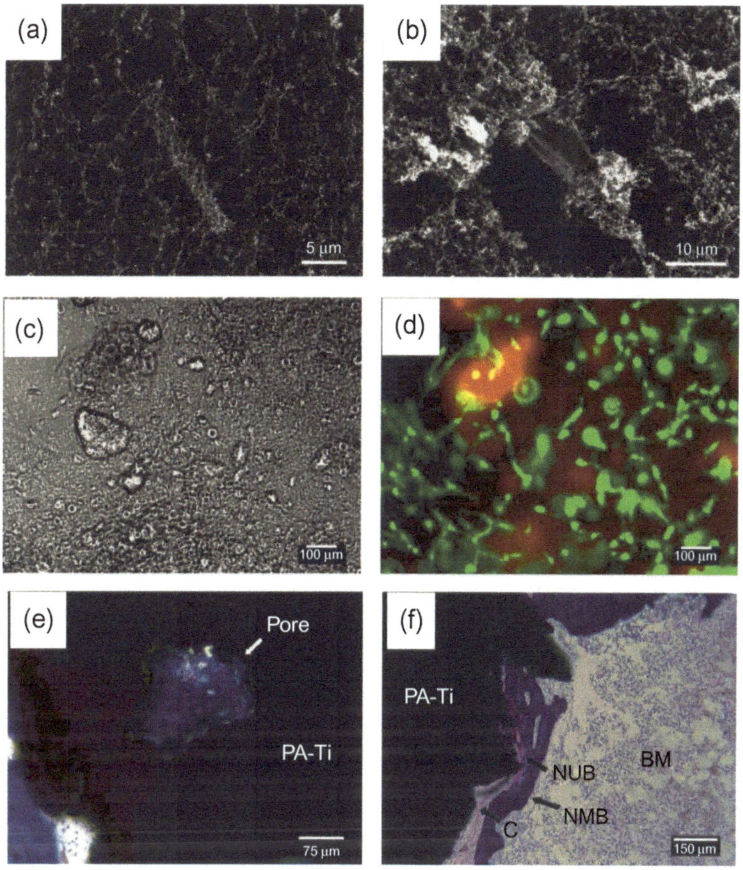

Figure 6.13 SEM images of (a) the surface of Ti-6Al-4V foam modified with PA self-assembled nanofibers and (b) the MC3T3-E1 preosteoblastic cells encapsulated near the surface of the PA gel, respectively. (c) Optical and (d) fluorescence microscopy images of MC3T3-E1 cells encapsulated within the nanofiber matrix of the PA, where green fluoresced cells are indicative of live cells. (e) and (f) are images of histological sections of a PA–Ti hybrid implanted in a rat femur after 4 weeks. Nondecalcified, plastic embedded samples were stained with methylene blue and basic fuchsin. Image (e) shows mineralized bone formation (blue) within an interior pore of the PA–Ti hybrid. Image (f) shows new bone formation adjacent to and into the PA–Ti hybrid. The location and formation of these bone spicules are evidence of osteoconduction, with new mineralized bone (NMB, deep blue) on the exterior, new unmineralized bone (NUB, pink) in the middle, and collagenous fibers (C) against the PA–Ti hybrid.
Modified with permission from Ref. [20].

demonstrated that the versatility of self-assembly is spurring their advancement to meet the requirements of bone tissue engineering [21].

Self-assembled nanostructures have a number of extraordinary properties, including bioactivity and biocompatibility, biodegradability and *in situ* assembly that allows for encapsulation of cells and delivery of drugs. Differentiated from other types of orthopedic materials, these properties in combination with hydrogel properties enable the possibility of using self-assembling nanostructures for minimally invasive surgery, and gene and cell therapies [22]. In addition, the last part of this chapter also emphasizes that self-assembled nanostructures can be decorated with various biologically active moieties and ligands, or have been used to functionalize existing biomaterials and implants, for drug delivery and bone tissue engineering purposes. In the near future, this strategy will be further elaborated for the possibility of creating distinct and better microenvironments for supporting regrowth of tissues [23].

Finally, the opportunities brought by the new processing or fabrication techniques such as 3-D printing and rapid prototyping are magnificent for self-assembled nanomaterials [1]. New processing and fabrication techniques that allow for the development of tissue engineering scaffolds with complicated geometry and microstructures beyond hydrogel would significantly extend the application of self-assembled nanostructures in the regenerative engineering of bone.

References

[1] Salgado AJ, Coutinho OP, Reis RL. Bone tissue engineering: state of the art and future trends. Macromol Biosci 2004;4:743–65.

[2] Rose FR, Oreffo RO. Bone tissue engineering: hope vs. hype. Biochem Biophys Res Commun 2002;292:1–7.

[3] Stephanopoulos N, Ortony JH, Stupp SI. Self-assembly for the synthesis of functional biomaterials. Acta Mater 2013;61:912–30.

[4] Whitesides GM, Grzybowski B. Self-assembly at all scales. Science 2002;295:2418–21.

[5] Viney C. Self-assembly as a route to fibrous materials: concepts, opportunities and challenges. Curr Opinion Solid State Mater Sci 2004;8:95–101.

[6] Mason JM, Arndt KM. Coiled coil domains: stability, specificity, and biological implications. ChemBioChem 2004;5:170–6.

[7] Scanlon S, Aggeli A. Self-assembling peptide nanotubes. Nano Today 2008;3:22–30.

[8] Adler-Abramovich L, Aronov D, Beker P, Yevnin M, Stempler S, Buzhansky L, et al. Self-assembled arrays of peptide nanotubes by vapour deposition. Nat Nano 2009;4:849–54.

[9] Matson JB, Zha RH, Stupp SI. Peptide self-assembly for crafting functional biological materials. Curr Opinion Solid State Mater Sci 2011;15:225–35.

[10] Rockwood DN, Preda RC, Yücel T, Wang X, Lovett ML, Kaplan DL. Materials fabrication from *Bombyx mori* silk fibroin. Nat Protoc 2011;6:1612–31.

[11] Altman GH, Diaz F, Jakuba C, Calabro T, Horan RL, Chen J, et al. Silk-based biomaterials. Biomaterials 2003;24:401–16.

[12] Foo CTWP, Lee JS, Mulyasasmita W, Parisi-Amon A, Heilshorn SC. Two-component protein-engineered physical hydrogels for cell encapsulation. Proc Natl Acad Sci 2009;106:22067–72.

[13] Webster TJ. Biomimetic helical rosette nanotubes and nanocrystalline hydroxyapatite coatings on titanium for improving orthopedic implants. Int J Nanomed 2008;3:323–33.

[14] Zhang L, Rakotondradany F, Myles AJ, Fenniri H, Webster TJ. Arginine-glycine-aspartic acid modified rosette nanotube-hydrogel composites for bone tissue engineering. Biomaterials 2009;30:1309–20.

[15] Mata A, Geng YB, Henrikson KJ, Aparicio C, Stock SR, Satcher RL, et al. Bone regeneration mediated by biomimetic mineralization of a nanofiber matrix. Biomaterials 2010;31:6004–12.
[16] Rajangam K, Behanna HA, Hui MJ, Han X, Hulvat JF, Lomasney JW, et al. Heparin binding nanostructures to promote growth of blood vessels. Nano Lett 2006;6:2086–90.
[17] Chow LW, Bitton R, Webber MJ, Carvajal D, Shull KR, Sharma AK, et al. A bioactive self-assembled membrane to promote angiogenesis. Biomaterials 2011;32:1574–82.
[18] Shah RN, Shah NA, Lim MMDR, Hsieh C, Nuber G, Stupp SI. Supramolecular design of self-assembling nanofibers for cartilage regeneration. Proc Natl Acad Sci 2010;107:3293–8.
[19] Hosseinkhani H, Hosseinkhani M, Tian F, Kobayashi H, Tabata Y. Bone regeneration on a collagen sponge self-assembled peptide-amphiphile nanofiber hybrid scaffold. Tissue Eng 2007;13:11–9.
[20] Sargeant TD, Guler MO, Oppenheimer SM, Mata A, Satcher RL, Dunand DC, et al. Hybrid bone implants: self-assembly of peptide amphiphile nanofibers within porous titanium. Biomaterials 2008;29:161–71.
[21] Wu EC, Zhang S, Hauser CA. Self-assembling peptides as cell-interactive scaffolds. Adv Funct Mater 2012;22:456–68.
[22] Beachley V, Wen X. Polymer nanofibrous structures: fabrication, biofunctionalization, and cell interactions. Prog Polym Sci 2010;35:868–92.
[23] Kyle S, Aggeli A, Ingham E, McPherson MJ. Production of self-assembling biomaterials for tissue engineering. Trends Biotechnol 2009;27:423–33.

Nanotechnology-controlled drug delivery for treating bone diseases

7

7.1 Nanotechnology and bone diseases

The interdisciplinary fields of nanobiotechnology, including nanomedicine, are revolutionizing the traditional disciplines of science. As a perfect example, research in the area of controlled drug delivery has experienced tremendous progress as of late thanks to the emergence of nanotechnology. Novel applications of nanomaterials are enabling site-specific drug delivery and controlled release of traditional pharmaceuticals, recombinant proteins, vaccines, and nuclei acids [1]. Novel drug delivery systems have been devised and fabricated by nanotechnology. These systems are approaching multifunctional "smart" entities with the ability to sense, diagnose, image, and cure numerous ailments [1–3]. Owing to these advancements, interest in nanotoxicity has also been soaring because it is widely unknown how the body reacts to drug delivery materials at this size range. While it may be true that the initial vision of nanotechnology toward enhancing drug delivery was the everlasting idea of a "nano-robot" or "nano-doctor" in which "nano-machines" continuously survey the body treating diseases on the spot, more practical applications of nanomaterials in drug delivery have emerged.

Among the broad spectrum of nanotechnology-based drug delivery applications, treating bone diseases is one of the active and urgent areas. There are several kinds of diseases associated with bone, most commonly osteoporosis, osteoarthritis, Paget's disease, and bone cancer [4]. In the United States today, 10 million people have osteoporosis and almost 34 million more have low enough bone mass to place them at an increased risk for developing osteoporosis [5]. Moreover, according to the American Academy of Orthopedic Surgeons, 19 million people visited a physician for osteoarthritic problems in 2006 [6]. It has also been estimated that 3% or 8.2 million people have Paget's disease currently in the United States. [7]. In addition, the American Cancer Society reported that about 2380 new cases of bone and joint cancer were diagnosed in 2008 and about 1470 deaths from these cancers are expected [8].

Current medications and clinical therapies for treating these bone diseases are summarized in Table 7.1. Most of the listed medications have been approved by the U.S. Food and Drug Administration (FDA). Although these medications and therapies are effective in preventing and treating bone diseases syndromes, most of these bone diseases are considered to largely uncurable [5,10]. Thus, it is clear that our current drug regimes for treating various orthopedic diseases need improvement. Besides the inherent side effects associated with all the medications in Table 7.1, delivering and retaining drugs specifically in bone (to increase their efficacy) is challenging owing to the specific composition, architecture, and structure of bone tissue limiting access of such drugs. Differing from other tissues, bone consists of 70% (dry weight) inorganic

Nanotechnology-Enhanced Orthopedic Materials. http://dx.doi.org/10.1016/B978-0-85709-844-3.00007-0

Table 7.1 **Current medications and clinical therapies for bone diseases**

Bone diseases	Current medications or therapies	Reference
Osteoporosis	Antiresorptive medications: bisphosphonates, calcitonin, estrogen and estrogen agonists/ antagonists; anabolic (bone forming) medications; parathyroid hormone; surgery or bone replacement	[5]
Osteoarthritis	Nonsteroidal anti-inflammatory drugs, cyclooxygenase-2 (COX-2) inhibitors, corticosteroids, and artificial joint fluid; surgery or bone replacement	[9]
Paget's disease	Risedronate, pamidronate, alendronate, zoledronic acid, etidronate, tiludronate, and calcitonin; surgery or bone replacement	[10]
Bone cancer	Surgery, radiation therapy, and chemotherapy	[11]

constitutes (including hydroxyapatite (HA) and tricalcium phosphate (TCP)) and related bone matrix proteins. Bone possesses one of the most delicate hierarchical micro- to nanostructures in the human body. These unique characteristics call for the proper selection of drug carriers that can target specific diseased parts of bone [12,13]. Similar to other fields, applying nanomaterials as effective and efficient drug delivery systems has provided a lot of promise to orthopedic applications. This is in large part due to the ability of nanoparticles to traverse the nanostructure of bone, attach to a site of bone disease, and prolong drug release [14]. Such efforts started with creating nanostructured ceramic–polymer composites as bone substitutes (without drug carriers) to mimic the exact constituent nanometer bone structure. Since these materials accelerated bone tissue regeneration, an obvious next step was the exploration of nanoparticles for drug delivery orthopedic applications [12–14].

The engagement of nanotechnology in drug delivery for treating various orthopedic diseases is new, but becomes important when considering bone regeneration using bone implants and replacement strategies. The material forms used for these drug delivery purposes span a wide range of nanoscale chemistries including (but not limited to) polymers, metals, ceramics, semiconductors, sol–gels, and self-assembled molecular complexes. Diverse nanomaterial geometries have also been fabricated including particles, fibers, capsules, tubes, whiskers, dendrimers, and so on. [1,15]. Such nanomaterials have been used to deliver both small molecular drugs and various classes of biomolecules tailored to specific-release kinetics and biodistribution [1], enabling effective remedies for first treating and, then second, enhancing bone regeneration by promoting bone cell growth. In many circumstances it is the orthopedic drug carrier itself that promotes new bone growth [12–14]. This is in extreme contrast to traditional drug delivery in which the function of the carrier is completed once the drug is released.

In some orthopedic nanotechnology drug delivery applications, the drug is released from the carrier to treat the disease but the carrier itself remains to stimulate new bone growth. Moreover, the drug carrier can be designed to provide immediate bone strength to the diseased bone before the drug is even released. In this chapter, the current developments and novel applications of orthopedic-related nano drug delivery are summarized and the trends of current studies and future directions outlined. However, it is necessary to emphasize that most of the nanomaterial-based drug delivery systems are still in the phases of research and experiments, and the understanding of their toxicity and safety is still lacking, thus, limiting clinical applications.

7.2 Nanomaterials as drug delivery systems: fundamentals and principles

A successful drug delivery system should be able to overcome the inherent limitation associated with biomacromolecular therapeutics (such as short plasma half-lives, poor stability, and potential immunogenicity) to maximize therapeutic activity while minimizing toxic side effects of the drug [1,16]. The key issues toward achieving these requirements include site-specific targeting, transportation of drug carriers, and the controlled release of drugs. Traditional drug delivery systems have difficulties in drug stability during transportation, prolonging drug release and releasing drugs into individual cells. These limitations (often times leading to an unhealthy initial burst release of drugs) can be more serious for drug delivery to bone than to other tissues. This is because when dealing with bone implants, a time delay and stable drug delivery with little initial burst release is desirable while the drug carrier maneuvers the complex intricacies of bone structures to get to the diseased site [17].

The inherent properties of nanoscale materials (such as physical, chemical, mechanical, electrical, magnetic, and optical properties) can be utilized to strengthen the performance of drug delivery systems. Specifically, the nanometer size of a drug carrier provides numerous advantages for drug delivery purposes, such as: (1) enhanced transport across cell membranes, thus, reducing clearance from the body and providing for more targeted drug delivery [1,16]; (2) greater surface area-to-volume ratios and subsequently more surface reactivity, thus, increasing drug loading ability, providing controlled dissolution rates, and drug bioavailability [18]; (3) increased dispersibility for homogeneous drug loading and release of drug molecules; (4) promoted mechanical properties (such as matching the strength, ductility, etc., of natural bone), thus, serving as a strong immediate matrix backbone; and (5) size similarity to natural tissue components (e.g., HA crystals in natural bone are 50×25 nm [19]), thus, enabling better tissue acceptance by biomimicking tissue architecture.

Several studies have reported that the size variation of drug carriers within the nanoscale range strongly affects drug bioavailability and blood circulation times [1,20,21]. For example, particles with diameters <10 nm can be rapidly removed through extravasation and renal clearance after systemic administration, while particles with sizes in the range of 70–200 nm revealed prolonged circulation times [1,20]. A wide size distribution of nanoparticulates can also vary gene transfection efficiency in gene delivery therapies [15]. Therefore, a preferred size range of 10–100 nm for nanomaterial drug carriers has been suggested [1].

In addition to the influence of nanometer particle size, a variety of geometries and architectures of drug carriers have been studied. As an example, nanoscale hollow structures (specifically, tubes, cages, shells, etc.) have demonstrated great control over drug loading amounts and time of release [22]. Specifically, by controlling tube length, diameter, and wall thickness, one can control many aspects of drug delivery (such as loading amounts and time of release [22]. Another study proposed a drug delivery system in which drugs were conjugating on gold nanocages and the nanocages simultaneously served as optical imaging contrast agents to tag cancer sites and release drugs [23]. In addition, distinctive magnetic, electrical, and optical properties of nanomaterials can also benefit controlled drug delivery processes by assisting in the external control of drug transportation, targeting, and even simultaneously sensing, diagnosing, and treating the disease [24–26]. Conventional drug delivery processes often only treat diseases, but do not play a role in diagnosing disease.

To construct better drug delivery systems through nanotechnology, surface properties of nanomaterials have been carefully tailored through physical or chemical adsorption methods. Usually, the physical or chemical adsorption of drugs to nanoparticles has been governed by the inherent chemistry or physical properties of the materials, which has also influenced drug loading, particle transportation, and drug release, as well as biocompatibility and degradability properties. Surface hydrophilicity of the drug carriers has also been a key surface property. Specifically, this property can determine the construction and assembly of nanomaterials *in situ*, affecting adsorption/desorption of drug and proteins, cell and tissue responses as well as clearance from the body. For example, several types of proteins (such as fibronectin and vitronectin) that promote osteoblast (bone-forming cells) functions are hydrophilic and, thus, prefer to adsorb on hydrophilic surfaces [27]. If these proteins adsorb in high quantities (as they have been shown to do on nanoparticles of ceramics, metals and polymers), bone growth ensues. Electrical charges on drug carriers have also been shown to be effective for drug loading. Specifically, electrical properties on surfaces interact strongly with drugs or biological systems to promote even greater impact on drug bioactivity, drug release kinetics, conjugation to targeting moieties, and transport in bone. For instance, cationic nanoparticles can localize in the cytoplasma and within mitochondria, while anionic nanoparticles remain in lysosomes [1,28]. Last, for bone regeneration, the porosity of nanomaterial drug carriers is another key factor that can be tailored to promote drug delivery [29]. Specifically, porosity of particles controls degradation, material solubility, and drug release profiles. Particle porosity also affects material biocompatibility properties and bone cell growth potential.

7.3 Nanoparticulate drug delivery systems

Owing to simple synthesis, high reproducibility and feasibility of less invasive administration, particulate materials have been widely investigated in numerous areas of drug delivery for treating bone diseases and reconstructing bone tissue (Figure 7.1). Differing from larger size drug-delivery particles that release high local drug concentrations, nanoparticulate drug delivery systems can be endocytosed directly and, thus, can

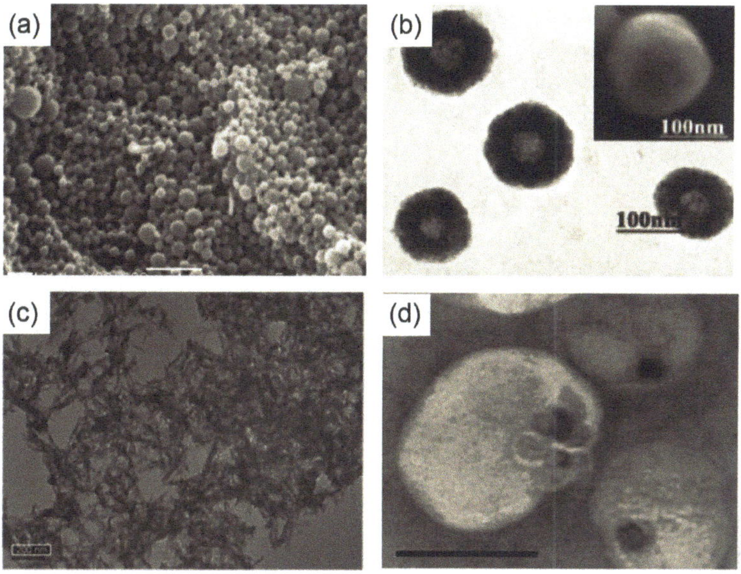

Figure 7.1 Representative nanoparticulate drug delivery systems for treating bone diseases. (a) Poly(lactic-co-glycolic acid) (PLGA) nanoparticles, scale bar = 1 μm; (b) Hollow calcium phosphate nanospheres, scale bar = 100 nm; (c) Calcium phosphate-coated γ-Fe_2O_3, scale bar = 200 nm; and (d) Magnetic liposomes, scale bar = 100 nm.
Reprinted from (a) Ref. [30], (b) Ref. [31], (c) Ref. [32], and (d) Ref. [33], with permission.

release drugs either outside or inside target cells (allowing for smaller amounts of drug delivery to achieve a desired effect [1]). Nanoparticulates also have benefits toward gene delivery therapy owing to their sizes, which are small enough to be internalized into the cell, thus, providing the capability of escaping endosomes/lysomes (which protect DNA) until they reach the target [34]. In addition to size benefits, nanoparticles also have various properties that can be tailored for specific applications owning to their exceptional bulk or surface properties. For example, in addition to their ability to bind peptides and proteins in a much more bioactive manner than micron particles, optical and photothermal properties of gold nanorods can be used to specifically target and image cancer cells by scattering light, and finally lysing cancer cells by converting light energy into heat [25]. The following sections provide more specific specific examples of the use of polymer, inorganic and composite nanoparticles in orthopedics.

7.3.1 Polymeric nanoparticles

Polymeric nanoparticles are probably the largest category of nanomaterials used in drug delivery owing to their easily tailored biodegradability properties. Drug loading and release profiles of polymeric nanoparticles can be easily adjusted by altering molecular weight (MW), surface hydrophilicity/charge, and free functional groups. Poly(D,L-lactic-co-glycolic acid) (PLGA) is a good example of such common drug delivery nanoparticle and has been modified for bone-specific drug delivery applications

(Figure 7.1a). PLGA has a highly hydrophilic surface with numerous carboxylate end groups. A method of controlling the density of reactive carboxyl groups on PLGA to prolong drug release was recently reported [35]. This study revealed that adjusting the combination of high MW encapped and low MW nonencapped PLGA could change the carboxyl density with a higher proportion of high-MW PLGA contributing to prolonged celecoxib (4-[5-(4-methylphenyl)-3-(trifluoromethyl)-1H-pyrazol-1-yl] benzene-sulfonamide commonly used for treating osteoarthritis and rheumatoid arthritis) release. Moreover, modified PLGA nanoparticles have exhibited great potential for orthopedic drug delivery [36]. For example, alendronate, a type of bisphosphonate that has strong affinity for binding to calcium phosphate in bone and has been used clinically for treating osteoporosis, can be conjugated to PLGA and collectively these structures are expected to have a great ability to target bone and release drugs at a specific site [37]. Besides albumin microparticles, micro- or nanoparticulate PLGA is also a possible candidate in intra-articular drug delivery in osteoarthritis [38]. Efficacy of intraarticular PLGA nanoparticles (300–490 nm) loaded with betamethasone phosphate can last about 6 weeks. However, issues such as fast cellular uptake and reproducibility still need to be investigated before launching the nanoparticulate PLGA-based drug systems for intra-articular applications.

Copolymers also exhibit great potential for bone-specific drug delivery because they can combine advantages of different polymers. This is also clearly true at the nanometer level [15,39]. One drug delivery strategy has been to create copolymers of different hydrophilicity properties to achieve self-assembled structures such as polymeric micelles (sizes about 50–100 nm) with a hydrophobic core and hydrophilic outer shell. Through physical or covalent bonding, hydrophobic cores can be used to contain hydrophobic drugs (such as antiphlogistic and anticancer drugs, DNA or other proteins [40–42]). As another example, a novel heparin-conjugated micelle has been shown to incorporate and deliver both basic fibroblast growth factor (bFGF, used as a model growth factor) and hydrophobic indomethacin (IMC, used to treat rheumatoid arthritis and gout) [43]. The micelle used Tetronic®-poly(ε-caprolactone) (PCL)-heparin as a copolymer and *in vitro* drug release profiles showed that IMC slowly released over 3 weeks while bFGF released over 2 months in a controlled manner. Coupling hydrophobic and electrostatic interactions, more complex micelle like core-shell structures were obtained. The detail and properties of these structures have been discussed in several recent reviews [44,45].

Another strategy centered on copolymerization is to use poly(ethylene glycol) (PEG) to increase blood circulation times of the drug carrier and mask bounded biomolecules (proteins, DNA, etc.), also known as PEGylation, until the carrier gets to its destination [15,46]. For example, PEG-PLGA nanoparticles have enhanced protein delivery and biodistribution in a rat by extending the half-life of bovine serum albumin (BSA) 20 times compared to BSA loaded in PLGA alone [47]. PEGylated copolymers (such as methxy poly(ethylene glycol)/poly(ε-caprolactone) (MPEG/PCL), amine-terminated MPEG/PCL, and PEG-cationized gelatin) have also been used as nonviral gene delivery carriers and for modulating DNA transfection efficiency, which will not be discussed in detail here due to space but can be found in several recent literature articles [48,49].

7.3.2 Inorganic nanoparticles

Owing to chemical similarities to bone, nanoparticulate inorganic materials have also been extensively studied for orthopedic drug delivery. In contrast to polymeric nanoparticles, inorganic particles have much longer biodegradation times or are even undegradable, which clearly alters drug release mechanisms and has a greater impact on subsequent drug release kinetics. In addition, inorganic nanoparticles exhibit electrical, mechanical, magnetic, and optical properties that are rarely seen in polymeric materials. These distinct properties have led to several novel applications in bone-specific drug delivery as reviewed here.

Calcium phosphate nanoparticles (including HA, TCP, calcium tetraphosphate, and their derivatives) have been widely studied and used in bone tissue engineering applications and bone-specific drug delivery due to their biocompatibility, bioabsorbability, and bioactivity. In addition, owning to their chemical, structural, and size similarity to the inorganic components of natural bones, calcium phosphate nanoparticles are perfect for bone regeneration purposes. Moreover, rich hydroxyl groups on HA as well as the hydrophilic property of calcium phosphates readily enable the conjugation of both small- and large-molecule drugs. Recently, hollow-structured calcium phosphate nanospheres were designed and fabricated to achieve an "on-off" release of drugs using ultrasound (Figure 7.1b) [31]. The novel HA nanospheres were fabricated by an ultrasonic-assisted wet-chemical reaction with hexadecyl (cetyl) trimethylammonium bromide (CTAB) as a modifier. These uniform and monodispersed hollow HA structures can collapse and transfer into pin-shaped HA nanocrystallites under ultrasonic treatment. In this manner, the drug release of encapsulated compounds can be triggered by ultrasound and regulated by altering ultrasound power density. An experiment using amylose indicated that the drug-released amounts could be consistently controlled by ultrasound application time [31,50]. Calcium phosphate nanoparticles are also potential candidates for bone-specific gene delivery. Numerous studies have revealed that nano-calcium phosphates possess a higher penetration rate into cell membranes and their transfection efficiency can be up to 25-fold higher than that of conventional particles. In addition, nano-HA can load greater amounts of drugs because of their notoriously high adsorption affinity and high area-volume ratio [50].

Magnetic nanoparticles have also attracted increasing interest in orthopedic drug delivery. The most widely studied magnetic nanoparticles are nanoscale iron oxides including magnetite γ-Fe_2O_3, magnetite Fe_3O_4, and associated compounds (also known as superparamagnetic iron oxide nanoparticles (SPION)), which have been actively used for magnetic resonance imaging, hyperthermic treatment, and non-bone drug delivery [51,52]. Recently, a strategy of using SPION to treat bone diseases (such as osteoporosis, osteoarthritis, and bone cancer) was proposed using biocompatible SPION conjugated with specific drug molecules, which were delivered to osteoporosis sites or diseased tissue in the presence of a directional magnetic field [32,53]. The drug molecules ranged from bone growth factors (such as BMP-2) to those for treating bone diseases (such as bis- and di-phosphonates and estrogen modulators). A preliminary investigation aimed at treating osteoporosis demonstrated that the iron oxide nanoparticles could be coated with calcium phosphate and then modified with bioactive agents (YRGDSPC) to promote bone growth while still retaining magnetic properties [53]. Another preliminary study revealed that γ-Fe_2O_3 nanoparticles and γ-Fe_2O_3 coated with calcium phosphate (Figure 7.1c) in the presence of BSA significantly increased osteoblast proliferation compared to controls without nanoparticles [32]. Such promise in the use of inorganic nanoparticles at promoting bone growth without using additional drugs is unprecedented. Coupled with the successful applications of SPION in biomedical imaging, SPION render drug delivery systems multifunctional for purposes of imaging, diagnosis, and therapy.

More specifically for bone cancer, novel selenium nanoparticles have been developed (again without the use of drugs) [54]. Selenium naturally exists in the human body and is known to reduce various types of cancers (including pancreatic, gastric, lung, nasopharyngeal, breast, uterine, respiratory, digestive, hematological, and gynecological) [55]. Biocompatibilty of selenium and the anti-carcinogenic effects of selenium nanoparticles synthesized through a colloidal technique have been evaluated [54,56,57]. Results showed that nanoparticles of selenium have bone-specific biocompatibility and when grown on titanium have significantly inhibited the competitive growth of cancerous osteoblasts while increasing the growth of healthy osteoblasts. Clearly, these new findings support the feasibility of using inorganic nanoparticles (as drug carriers) for bone cancer treatments.

7.3.3 Nanoparticle composites

As mentioned in the previous section, polymeric nanoparticles are biodegradable and biocompatible, while inorganic nanoparticles possess many distinctive physical properties. Harnessing these properties into a single composite drug delivery system enables a synergic and multipurpose therapy for treating bone diseases. As an example, nanomagnetic liposomes (Figure 7.1d) fabricated by combining magnetic moieties with phospholipids have been used as carriers for recombinant human bone morphogenetic protein-2 (rhBMP-2) for bone regeneration in a rat bone defect [33,58]. The results showed that a single topical application of magnetic liposomes with an appropriate amount of rhBMP-2 (3 μg) incorporated under magnetic induction immediately after surgery was effective at promoting new bone formation. Moreover, the combined treatment of topical magnetic rhBMP-2 liposomes and magnetic force assisted implantation was effective for healing bone defects [33]. A similar rationale of achieving desirable properties by combining corresponding nanoparticles has been widely applied in orthopedics, and other examples are not discussed here due to space limitations but can be found in several recent reviews [15,24,59].

7.4 Nanostructured scaffolds as controlled release matrices

Nanostructured scaffolds can serve multiple purposes including supporting cellular functions by mimicking biological micro- and nanoenvironments (such as the bone extracellular matrix (ECM)) and controlling drug release. Scaffold-based drug delivery can be achieved by the incorporation of signaling molecules into scaffolds through simple dispersion or through stronger immobilization to nanoparticles by electrostatic or covalent bonding [60]. In such molecule-incorporated and nanoparticle-derived scaffolds, besides bulk material or surface properties, the architecture (such as hierarchical structures, alignment, assembly of the building block materials and associated porosity have been adjusted to play key roles in drug release kinetics, drug stability, drug transportation, scaffold degradation, and biocompatibility. Additionally, because of their structure and volume, scaffolds are typically delivered by invasive measures, which significantly differ from the less invasive routes of administration for particulate drug delivery systems.

7.4.1 2-D nanostructured surfaces

2-D nanostructured surfaces refer to coatings, films, or other layers with nanoscale features and topographies, which have much larger scales in two of the three dimensions than the third one (Figure 7.2). These surfaces are usually attached or coated on substrates or other bulk materials such as orthopedic implants to provide biologically active surfaces for drug delivery. 2-D nanostructured surfaces usually receive less attention than 3-D scaffolds for tissue engineering but could play a very important role for currently used metallic or ceramic implant efficacy. In contrast to nanoparticles, 2-D nanostructured surfaces (including surface roughness, topography, surface energy, and surface chemistry) have a much more profound impact on controlling protein

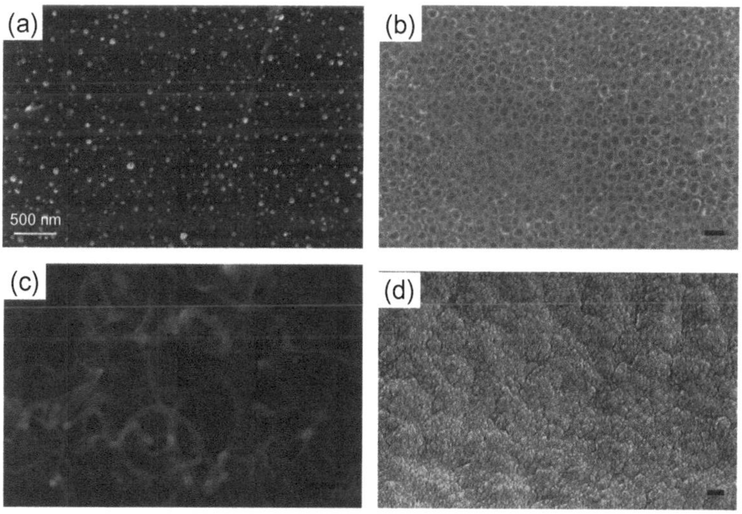

Figure 7.2 Representative 2-D nanostructured surfaces for bone-specific drug delivery systems. (a) Selenium nanoparticle-coated titanium surface, scale bar = 500 nm (reprinted from Ref. [54] with permission); (b) nanotubular anodized titanium surface, scale bar = 100 nm (reprinted from Ref. [26] with permission); (c) carbon nanotubes (CNT) grown out of anodized nanotubular titanium surface, scale bar = 200 nm (reprinted from Ref. [26] with permission); and (d) nanocrystalline diamond surface, scale bar = 200 nm.

adsorption, subsequent cellular events, and of course, drug release directly to adjacent bone [61–64].

For example, the selenium nanoparticles mentioned in the previous section have been coated on titanium and stainless steel implants to provide an anti-carcinogenic coating (Figure 7.2a) [65]. This study showed that selenium nanoclusters could be firmly bonded and well dispersed on titanium and stainless steel implant surfaces through a simple colloidal technique. The payload of selenium on implants can be controlled by the concentration of selenium in the synthesis solution. Results revealed an increase in noncancerous osteoblast density on selenium-coated titanium compared to uncoated samples and a similar osteoblast density on both coated and uncoated stainless steel (both after 4 hours and 1 day). These results suggested that selenium-coated implants are cytocompatible, but more importantly, the function of cancerous osteoblasts on these selenium-coated surfaces was significantly inhibited compared to results obtained on uncoated materials for up to 3 days. This inhibiting effect was also enhanced as the selenium concentration on the implants increased.

Besides treating bone cancer, many 2-D nanostructured drug carriers have been designed to promote bone regeneration. Specifically, nanotubular anodized titania surfaces with a tube width of tens of nanometers and a tube length of a few hundred nanometers have been promising bone-growing candidates (Figure 7.2b). Nanotubular titania surfaces can be readily fabricated by direct anodization of titanium implants [66,67]. Anodization, as introduced in previous chapters, is a technique to grow an *in situ* metallic oxide layer on metal surfaces in an electrochemical cell that uses the metal as an anode and frequently uses platinum as a cathode (as well as a water-based electrolyte solution). To create nanoscale tubular or porous topographies on newly formed metallic oxide layers, a voltage between 10 and 40 V and fluorine-based electrolytes are often used. The mechanisms of forming nanometer features through anodization are still under investigation but, usually, the local electric sparking at defect sites on the dielectric metallic oxide layer and electric field-assisted dissolution of the metallic oxide layer are believed to account for the formation of nanoscale features [67,68]. Theoretically, anodization can be applied to any metal that is stable to oxide to fabricate nanoscale tubular or porous surfaces. The unique tubular structures in metals can serve as a source of loading bone growth factors or antibiotics to promote bone growth or suppress bacterial infection when released [67–70]. This is a novel application for metals because anodization can transform a metal into a drug delivery device. Importantly, metals are the most common orthopedic implant material and yet, few metallic drug delivery strategies exist from metals.

A recent study has shown that immobilizing the amino acid peptide sequence of a segment of BMP-2 (the knuckle epitope, CKIPKASSVPTELSAISTLYL) into anodized titanium nanotubes (approximately 70 nm in diameter and 200 nm in depth) can promote osteoblast adhesion [70]. Another study modified anodized titanium nanotubes to increase drug-loading efficiency and prolong drug release by attaching amine or methyl groups to the anodized nanotubular titanium surfaces via a silanization process [71]. Results of that study revealed a prolonged continuous release of antibiotics and anti-infection agents decreasing *Staphylococcus epidermidis* (*S. epidermidis*) colonization [71]. A similar result of incorporating titania nanotubes with gentamicin (an aminoglycoside antibiotic particularly useful for treating gram-negative infection) also significantly reduced initial *S. epidermidis* adhesion on orthopedic implants [71]. In addition, immersing nanotubular titania into a mixture of simulated body fluid (SBF) and penicillin G created a drug-loaded calcium phosphate coating that prolonged drug release for up to 21 days. These results clearly indicated the promise of using metallic nanotubular surfaces for controlling drug release and improving bone implant efficacy.

More promise exists for the controlled drug release from orthopedic implants by using novel self-assembled material coatings. Specifically, helical rosette nanotubes (HRN), soft organic materials obtained through the self-assembly of low MW synthetic DNA molecules, have been developed to mimic the natural nanostructure of collagen and other components in bone [72]. HRN have demonstrated an exceptional potential to enhance osteoblast adhesion when either coated on titanium or incorporated into HA [73]. One recent study focused on conjugating arginine-glycine-aspartic acid (RGD) peptides (a cell adhesive peptide) or lysine on HRN and placing this new material on solidified polymerized 2-hydroxyethyl methacrylate (pHEMA) hydrogels [74]. Results indicated up to a twofold enhancement in osteoblast adhesion on 10% HRN-RGD-K (HRN loaded with RGD and lysine) coated hydrogels than on uncoated hydrogels and 73% more than on collagen-coated hydrogels of similar concentration. Fibronectin adsorption tests also revealed increased protein adsorption on RGD or lysine conjugated HRN. A recent study also showed that HRN loaded with dexamethasone prolonged drug releasing for up to 28 days due to the penetration of the drug down the long (up to several microns) interior of the HRN tube. These results highlight the promise that self-assembled bioactive systems have to deliver specific protein segments or molecules to promote bone growth.

Carbon nanotubes (CNT) or carbon nanofibers (CNF) have also been studied as a local drug delivery system to promote and guide bone-tissue regeneration [75,76]. CNT have been successfully coated or patterned on titanium, anodized nanotubular titanium (Figure 7.2c), and polycarbonate urethane [77,78]. A recent animal study revealed that multiwalled CNT combined with collagen and rhBMP-2 accelerated bone formation after implantation in a mouse muscle [79]. This finding is important toward the development of new drug delivery systems or scaffold materials for bone regeneration using multiwalled CNT as it also demonstrated no toxicity. Furthermore, rhBMP-2 loaded multiwalled CNT can be coated on implants (such as plates and screws) to form 2-D nanostructured surfaces for rapid bone healing.

Lastly, nanocrystalline diamond (NCD) has been studied as a versatile 2-D nanostructured surface for orthopedic drug release purposes (Figure 7.2d). Specifically, BMP-2 has been immobilized on oxygen-terminated NCD by physisorption and the drug-coated implant enhanced *in vivo* osseointegration when inserted into sheep calvaria [72–80]. More excitingly, this BMP-2 coated NCD system strongly activated the expression of osteogenic markers in mesenchymal stromal cells (MSC) cultured on this surface. In addition to superior antiwear properties and knowledge that the topography of NCD exhibits greater cytocompatibility properties than diamond films of larger grain sizes (i.e., submicro- or microcrystalline diamond) [81,82], an atypical multifunctional drug delivery platform to promote or inhibit bone growth can be created, depending on what events are desired per implant location.

7.4.2 3-D nanostructured scaffolds

3-D drug delivery scaffolds possess a greater resemblance to bone compared to the aforementioned 1-D or 2-D structures. Different from most 2-D surfaces that serve as coatings on implants, most 3-D scaffolds are biologically degradable or active and possess cell-inductive architectures (such high porosity), enabling the enhanced recruitment and migration of peripheral host cells into the scaffolds. Therefore, 3-D nanostructured scaffolds are more versatile and ready for controlled drug delivery to promote cellular events or guide morphogenetic processes [60,83,84]. In this case, 3-D scaffolds serve both as a temporary scaffold for promoting cell activity and as a delivery vehicle for controlling drug release, which is more suitable for bone regeneration after removal of diseased bone tissues [85]. Here, recent progress of bone-specific drug delivery using 3-D scaffolds is reviewed, focusing on those inspired by nanotechnology (Figure 7.3).

Electrospun nanostructured scaffolds have been intensively investigated as 3-D scaffolds for bone engineering. The principle of electrospinning is to apply high voltage on polymer solution or melt to overcome its surface tension and draw the

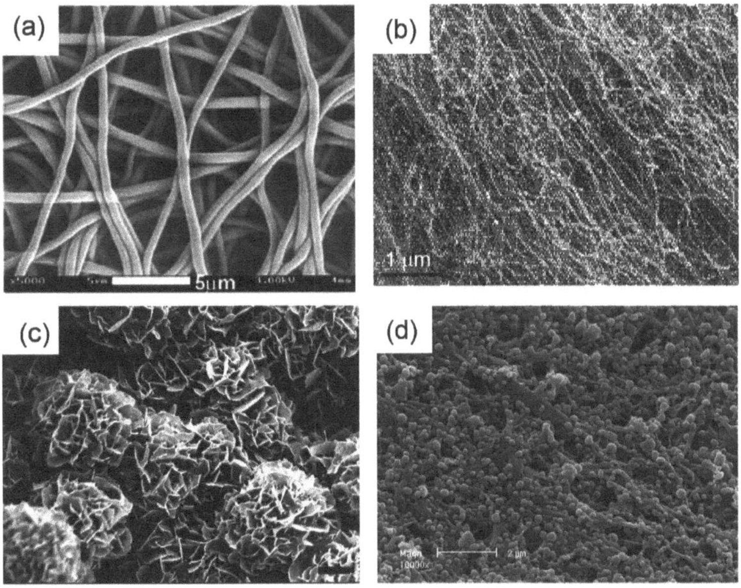

Figure 7.3 Representative 3-D nanostructured scaffolds for bone-specific drug delivery systems. (a) Electrospun silk scaffold with BMP-2 loaded, scale bar = 5 μm (reprinted from Ref. [86] with permission); (b) Self-assembled peptide-amphiphile (PA) nanofibers network, scale bar = 1 μm (reprinted from Ref. [87] with permission); (c) Nanocrystalline apatite modified poly(lactide-co-glycolide) (PLAGA) microsphere scaffolds, scale bar = 2 μm (reprinted from Ref. [88] with permission); and (d) poly(L-lactic acid) (PLLA) nanofibrous scaffolds incorporated with poly(lactic-co-glycolic acid) (PLGA) nanospheres, scale bar = 2 μm (reprinted from Ref. [89] with permission).

charged solution through an electric field from a nozzle onto a collector plate to form 3-D (or 2-D) structures. Due to the ease of fabrication, ability to control final product structure and extreme versatility in terms of the types of materials that can be used (mainly polymers), electrospinning has been widely used to prepare porous fibrous scaffolds with tunable drug release profiles and biodegradabilities. Advantages of electrospun scaffolds include high porosity, high surface-to-volume ratio, and a structual similarity to fibrous proteins in native extracellular matrices or collagen fibrils in bone. [1,15]. As drug-loaded platforms, electrospun PCL 3-D scaffolds carrying the drug simvastain were recently studied for the performance of regenerating bone. A rat study recently revealed significant osseous integration in the simvastain-loaded PCL scaffolds and better *in vivo* bone mineralization compared to the PCL scaffolds alone after 6 months [90]. Natural polymers (such as silk) have also been electrospun into scaffolds and loaded with molecules such as BMP-2 to promote human bone marrow-derived mesenchymal stem cells growth and differentiation toward osteogenic outcomes (Figure 7.3a) [86]. *In vitro* results showed that the BMP-2 loaded silk scaffolds supported higher calcium deposition and enhanced transcript levels of bone-specific markers than the silk scaffolds alone, indicating that the silk scaffolds were excellent vehicles for BMP-2.

Injectable nanoapatite scaffolds have also been developed for cell/growth factor delivery [91]. Tetracalcium phosphate ($Ca_4(PO_4)_2O$) and dicalcium phosphate ($CaHPO_4$) have been fabricated into self-setting calcium phosphate cements (CPC), and CPC have been further synthesized into strong nanostructured scaffolds using chitosan and various porogens (such as absorbable fibers and mannitol). In particular, water-soluble mannitol crystals which dissolve in a day and absorbable fibers which degrade after weeks, even months, have been used to create macropores in CPC. Clearly, it has been hypothesized that new bone could gradually grow into the macropores formed sequentially by mannitol and fibers after implantation, thus, maintaining the strength of scaffolds during bone growth. Studies have shown good attachment and proliferation of MC3T3-E1 osteoblast-like cells on CPC infiltrating into the macropores and anchoring into nanostructured apatite crystals on the pore walls. This novel scaffold was also modified to control the release of growth factors by adjusting liquid and chitosan ratios.

Bone regeneration though the controlled release of BMP-2 from an injectable peptide amphilphile (PA) nanofibrous scaffold has also been observed [87]. 3-D nanofibrous scaffolds were formed from the self-assembly of PA in the presence of BMP-2 (Figure 7.3b). *In vivo*, the BMP-2 release profile showed a 24-day release from PA nanofibrous scaffolds after the scaffolds were implanted into the back subcutis of a rat, compared to a 2-day duration of direct administration. This elongated release of BMP-2 induced significant homogeneous ectopic bone formation around the injected site.

Several 3-D nanocomposites have been designed to impart bioactivity, controlled drug loading, and drug release. For example, poly(lactide-co-glycolide) sintered microsphere scaffolds were modified by growing nanocrystalline apatite on their surfaces to enhance drug adsorption and slow drug release (Figure 7.3c) [88].

Heparin-conjugated PLGA scaffolds have been fabricated though techniques such as gas-foaming/salt-leaching and have been used to sustain drug delivery of growth factors like BMP-2 to enhance bone formation [92,93].

Lastly, several recent studies have focused on the subtle tuning of bone-specific drug release profiles rather than simply prolonging release time. For example, sequential growth factor delivery from 3-D scaffolds was realized using microspheres of polyelectrolyte complexes of alginic acid and poly(4-vinyl pyridine) [94]. The microspheres were loaded into PLGA scaffolds in the presence of growth factors (BMP-2 and BMP-7) and their effects on bone marrow stem cell growth studied. Results indicated that the sequential release of BMP-2 and BMP-7 enhanced osteogenic differentiation. In order to control drug release in a temporal fashion, a novel system of incorporating nanospheres into prefabricated nanofibrous scaffolds (NS-scaffolds) was developed (Figure 7.3d) [89,95]. *In vitro* release kinetics indicated that NS-scaffolds could release BMP-7 in a temporally controlled manner, exhibiting tunable release phases (especially, a controlled initial burst release) depending on chemical and degradation properties of the nanospheres. *In vivo* test results showed that the NS-scaffolds actively induced new bone formation throughout the scaffold as a result of BMP-7 release.

7.5 Summary and future perspectives

Recent research has demonstrated that nanotechnology can assist drug delivery and have exceptional potential for treating various bone diseases owing to their ability to modulate drug release kinetics, incorporate multifunctional molecules, target specific bone sites or cells, and respond to various external signaling sources (e.g., biological, electric, magnetic, or mechanical signals). These novel drug delivery systems involve an extensive spectrum of materials (from polymers to ceramics and from 1-D to 3-D) and a plethora of fabricating techniques that will continuously revolutionize the orthopedic drug delivery area. Although most of the *in vitro* or *in vivo* studies reviewed here are still distinct from clinical trials or human applications, and toxicity of nanomaterials is still an increasing concern, the extraordinary properties of nanomaterials mentioned here and the continual understanding of physical and biological properties of bone at the nanometer scale will open exciting and promising avenues to diagnose, understand, and treat numerous bone diseases.

In this chapter, the focus is placed on the latest experimental progression in controlling drug release for treating bone diseases and regenerating new bone in specifically desired locations. In every form of bone-specific drug delivery discussed (e.g., in applications involving nanoparticles, 2-D or 3-D nanostructured drug carriers), novel drug species, drug delivery kinetics, biocompatibility, and so forth have been optimized and improved using nanotechnology. Strong advances have been made by combining the advantageous properties of different forms of nanomaterials and by fabricating composites. This progress has allowed better drug delivery systems with tunable release profiles, adjustable degradability, promoted biocompatibility, and drug bioactivity with multiple functionalities. In addition to the strategy of creating composites discussed in previous sections, there are several ongoing critical topics or future directions in nanotechnology-derived controlled drug delivery.

The first topic is understanding of metabolism, toxicity and elimination routes of nanotechnology-derived drug delivery systems. Not just for bone diseases, but the whole nanotechnology-based drug delivery field lacks a fundamental understanding of how tissues and cells survive and function in the presence of nanomaterials. This has been a significant hurdle translating promising *in vitro* and animal studies to human clinic trials. An understanding of metabolism, toxicity and safety issues (such as manufacturing of nanomaterials) can provide essential guidance for the chemical and structural design, biodegradability, release kinetics and biocompatibility properties of nano drug delivery systems.

The second topic is refining drug delivery systems to achieve controllable multiphase drug release kinetics. In addition to the previous examples of delayed drug delivery, sequential delivery, and temporally controlled drug delivery, more complicated release mechanisms are being developed. These multiphasic drug deliveries include pulsed drug delivery, self-regulated drug delivery, and externally-regulated drug delivery [60]. Pulsed drug delivery, which releases drugs over specific periods of time, was used in PLGA-based systems but has not been used for treating bone diseases [100,101]. Self-regulated drug delivery is the one in which drugs are released as modulated in response to feedback information without external intervention. The biological feedback information that triggers release includes pH, biomolecule concentration, or existence of specific molecules [102–104]. In contrast, externally regulated drug delivery systems release drugs by an external stimulus including temperature difference, electrical and magnetic stimuli, as well as ultrasound [31,103–106]. These "smart" release mechanisms have not been widely used for orthopedic-specific treatments, but would be clearly beneficial for treating bone diseases and stimulating new bone growth. One recent attempt toward this approach is incorporating drugs into multiwalled CNT grown out of nanotubular titania and controlling drug release from these tubes through an electrical field.

Thirdly, gene delivery and multifactor delivery are future directions of great interests. Until now, drug molecules used for bone-specific treatments are mostly BMPs for bone growth and antibiotics for fighting bacterial infection. However, gene delivery is also a potent candidate for bone-specific drug delivery [107,108]. Furthermore, multifactor delivery in the same drug system could be an optimal solution for treating bone diseases and regulating tissue generation during different stages of bone repair [109,110]. One approach to achieve multifactor delivery is incorporating nanospheres into scaffolds [95]. Nanospheres would allow for a wider number of drugs and release profiles. Multiple drugs can be immobilized on different nanospheres of varied degradation rates, and in this manner, nanosphere-scaffold complexes can simultaneously deliver a variety of factors with varied release profiles.

Lastly, multifunctional scaffolds are also attractive platforms for treating bone diseases. As discussed before, 2-D and 3-D scaffolds can serve both as a cell/tissue support and a drug carrier. Recent research has attempted to extend the scope of scaffold functions and fabricate "smart" scaffolds that can simultaneously incorporate disease sensing, diagnosis, imaging, and treatment. To realize this, numerous nanomaterials can be used to fulfill these functions and immobilized into/onto scaffolds. For example, multiwalled CNT grown out of nanotubular titania have revealed the feasibility to support and sense new bone growth, scar tissue, and biofilm formation by measuring tissue conductivity [26]. Together with gene delivery [79,111], a "smart system" can be designed to stimulate new bone growth in response to an electrical feedback signal of lack of bone formation. Similarly, such materials can sense whether infection or inflammation is occurring by measuring such tissue conductivity, and drugs can then be released on demand. Nanotechnology is playing a key role in this process by measuring the conductivity of individual cells and proteins. As another example, magnetic nanostructures (e.g., SPION) with their ability to be used in magnetic resonance imagining (MRI) and thermal ablation, can be devised to image, target, and lyse bone cancer cells while at the same time promoting healthy new bone growth through growth factor delivery [112]. Some of these topics will be further discussed in Chapter 8.

References

[1] Goldberg M, Langer R, Jia X. Nanostructured materials for applications in drug delivery and tissue engineering. J Biomater Sci Polym Ed 2007;18:241–68.

[2] Anderson DG, Burdick JA, Langer R. Smart biomaterials. Science 2004;305:1923–4.

[3] LaVan DA, McGuire T, Langer R. Small-scale systems for in vivo drug delivery. Nat Biotechnol 2003;21:1184–91.

[4] Bone diseases, Medline Plus. A service of the US National Library of Medicine and the National Institutes of Health. Available at: http://www.nlm.nih.gov/medlineplus/bonediseases.html#cat22 [last accessed 16.02.09].

[5] National Osteoporosis Foundation. Available at http://www.nof.org/osteoporosis/diseasefacts.htm and http://www.nof.org/patientinfo/medications.htm [last accessed 16.04.09].

[6] American Academy of Orthopedic Surgeons (AAOS). Available at: http://www.aaos.org/Research/stats/Arthritis.pdf [last accessed 16.02.09].

[7] Prevalence and incidence of Paget's disease of bone. Available at: http://www.wrongdiagnosis.com/p/pagetbone/prevalence.htm [last accessed 16.02.09].

[8] What are the key statistics about bone cancer, American Cancer Society. Available at: http://www.cancer.org/docroot/cri/content/cri_2_4_1x_what_are_the_key_statistics_for_bone_cancer_2.asp? [last accessed 16.02.09].

[9] Glass GG. Osteoarthritis. Dis Mon 2006;52:343–62.

[10] Paget's Disease of Bone. Available at: http://www.niams.nih.gov/Health_Info/Bone/Pagets/medical_treatment.asp [last accessed 16.04.09].

[11] Bone tumor. Available at: http://orthoinfo.aaos.org/topic.cfm?topic=A00074&return_link=0 [last accessed 16.04.09].

[12] Ishizaki J, Waki Y, Takahashi-Nishioka T, Selective drug delivery to bone using acidic oligopeptites. J Bone Miner Metab 2009;27:1–8.

[13] Ezra A, Golomb G. Administration routes and delivery systems of biosphosphonates for the treatment of bone resorption. Adv Drug Deliv Rev 2000;42:145–95.

[14] Anitua E, Sanchez M, Orive G. Delivering growth factors for therapeutics. Trends Pharmacol Sci 2007;29:37–41.

[15] Kim K, Fisher JP. Nanoparticle technology in bone tissue engineering. J Drug Target 2007;15:241–52.

[16] Qiu LY, Bae YH. Polymer architecture and drug delivery. Pharm Res 2006;23:1–30.

[17] Gonzalez B, Colilla M, Vallet-Regi M. Time-delayed release of bioencapsulates: a novel controlled delivery concept for bone implant technologies. Chem Mater 2008;20:4826–34.
[18] Rabinow BE. Nanosuspensions in drug delivery. Nat Rev Drug Discov 2004;3:785–96.
[19] Liu H, Webster TJ. Bioinspired nanocomposites for orthopedic applications. In: Webster TJ, editor. Nanotechnology for the regeneration of hard and soft tissues. Singapore: World Scientific; 2007. p. 1–52.
[20] Ishida O, Maruyama K, Sasaki K. Size-dependent extravasation and interstitial localization of polyethyleneglycol liposomes in solid tumor-bearing mice. Int J Pharm 1999;190:49–56.
[21] Kong G, Braun RD, Dewhirst MW. Hyperthermia enables tumor-specific nanoparticle delivery: effect of particle size. Cancer Res 2000;60:4440–5.
[22] Popat KC, Eltgroth M, LaTempa TJ. Titania nanotubes: a novel platform for drug-eluting coatings for medical implants? Small 2007;3:1878–81.
[23] Sara ES, Chen J, Au L. Gold nanocages for biomedical applications. Adv Mater 2007;19:3177–84.
[24] Liu TY, Hu SH, Liu DM. Biomedical nanoparticle carriers with combined thermal and magnetic responses. Nano Today 2009;4:52–65.
[25] Huang XH, El-Sayed IH, Qian W. Cancer cell imaging and photothermal therapy in the near-infrared region by using gold nanorods. J Am Chem Soc 2006;128:2115–20.
[26] Sirivisoot S, Webster TJ. Multiwalled carbon nanotubes enhance electrochemical properties of titanium to determine in situ bone formation. Nanotechnology 2008; 19:295101.
[27] Wilson CJ, Clegg RE, Leavesley DI, Pearcy MJ. Mediation of biomaterial–cell interactions by adsorbed proteins: a review. Tissue Eng 2005;11:1–18.
[28] Panyam J, Labhasetwar V. Biodegradable nanoparticles for drug and gene delivery to cells and tissue. Adv Drug Deliv Rev 2003;55:329–47.
[29] Kujala S, Ryhanen J, Danilov A. Effect of porosity on the osteointegration and bone ingrowth of a weight-bearing nickel-titanium bone graft substitute. Biomaterials 2003;24:4691–7.
[30] Win KY, Feng S. Effects of particle size and surface coating on cellular uptake of polymeric nanoparticles for oral delivery of anticancer drugs. Biomaterials 2005;26:2713–22.
[31] Cai Y, Pan H, Xu X. Ultrasonic controlled morphology transformation of hollow calcium phosphate nanospheres: a smart and biocompatible drug release system. Chem Mater 2007;19:3081–3.
[32] Pareta RA, Taylor E, Webster TJ. Increased osteoblast density in the presence of novel calcium phosphate coated magnetic nanoparticles. Nanotechnology 2008;19:265101.
[33] Matsuo T, Sugita T, Kubo T. Injectable magnetic liposomes as a novel carrier of recombinant human BMP-2 for bone formation in a rat bone-defect model. J Biomed Mater Res A 2003;66A:747–54.
[34] Mansouri S, Cuie Y, Winnik F. Characterization of folate-chitosan-DNA nanoparticles for gene therapy. Biomaterials 2004;27:2060–5.
[35] McCarron PA, Marouf WM, Donnelly RF. Enhanced surface attachment of protein-type targeting ligands to poly(lactide-co-glycolide) nanoparticles using variable expression of polymeric acid functionality. J Biomed Mater Res 2008;87A:873–84.
[36] Choi SW, Kim WS, Kim JH. Surface-functionalized nanoparticles for controlled drug delivery. Methods Mol Biol 2005;303:121–31.
[37] Choi SK, Kim JH. Design of surface-modified poly(D, L-lactide-co-glycolide) nanoparticles for targeted drug delivery to bone. J Control Release 2007;122:24–30.

[38] Gerwin N, Hops C, Lucke A. Intraarticular drug delivery in osteoarthritis. Adv Drug Deliv Rev 2006;58:226–42.
[39] Wang D, Miller SC, Kopeckova P, Kopecek J. Bone-targeting macromolecular therapeutics. Adv Drug Deliv Rev 2005;57:1049–76.
[40] Kataoka K, Harada A, Nagasaki Y. Block copolymer micelles for drug delivery: design, characterization and biological significance. Adv Drug Deliv Rev 2001;47:113–31.
[41] Liu H, Farrell S, Uhrich K. Drug release characteristics of unimolecular polymeric micelles. J Control Release 2000;68:167–74.
[42] Riess G. Micellization of block copolymers. Prog Polym Sci 2003;28:1107–70.
[43] Lee JS, Bae JW, Joung YK. Controlled dual release of basic fibroblast growth factor and indomethacin from heparin-conjugated polymeric micelle. Int J Pharm 2008;346:57–63.
[44] Nishiyama N, Kataoka K. Current state, achievement, and future prospects of polymeric micelles as nanocarriers for drug and gene delivery. Pharmacol Ther 2006;112:630–48.
[45] Nishiyama N, Kataoka K. Nanostructured devices based on block copolymer assemblies for drug delivery: designing structures for enhanced drug formation. Adv Polym Sci 2006;193:67–101.
[46] Putman D. Polymers for gene delivery across length scales. Nat Mater 2006;5:439–51.
[47] Li Y, Pei Y, Zhang X. PEGylated PLGA nanoparticles as protein carrier: synthesis, preparation and biodistribution in rats. J Control Release 2001;71:203–11.
[48] Jang JS, Kim SY, Lee SB. Poly(ethylene glycol)/poly(ε-caprolactone) diblock copolymeric nanoparticles for non-viral gene delivery: the role of charge groups and molecular weight in particle formation, cytotoxicity, and transfection. J Control Release 2006;113:173–82.
[49] Kushibiki T, Tabata Y. Preparation of poly(ethylene glycol) induced cationized gelatin as a non-vial gene carrier. J Biomater Sci Polym Ed 2005;16:1447–61.
[50] Cai Y, Tang R. Calcium phosphate nanoparticles in biomineralization and biomaterials. J Mater Chem 2008;18:3775–87.
[51] Berry CC, Curtis ASG. Functionalisation of magnetic nanoparticles for applications in biomedicine. J Phys D Appl Phys 2003;36:R198–206.
[52] Xu C, Sun S. Monodisperse magnetic nanoparticles for biomedical applications. Polym Int 2007;56:821–6.
[53] Balasundaram G, Webster TJ. Applications of magnetic nanoparticles for the treatment of osteoporosis. Mater Res Soc Symp Proc 2007;1019:1–6, 1019–FF02–06.
[54] Tran P, Sarin L, Hurt RH. Titanium surfaces with adherent selenium nanoclusters as a novel anti-cancer orthopedic material. J Biomed Mater Res A 2010;93:1417–28.
[55] Navarro-Alarcon M, Lopez-Martinez MC. Essentiality of selenium in the human body: relationship with different diseases. Sci Total Environ 2000;249:347–71.
[56] Kopeikin VV, Valueva SV, Kipper AI. Synthesis of selenium nanoparticles in aqueous solutions of poly(vinylpyrrolidone) and morphological characteristics of the related nanocomposites. Polym Sci Ser A 2003;45:374–9.
[57] Tran P, Webster TJ. Enhanced osteoblast adhesion on nanostructured selenium compacts for anti-cancer orthopedic applications. Int J Nanomedicine 2008;3:391–6.
[58] Nawroth T, Rusp M, May RP. Magnetic liposomes and entrapping: time-resolved neutron scattering TR-SANS and electron microscopy. Phys B Condens Matter 2004;350:E635–8.
[59] Choi SK, Kim WS, Kim JH. Surface modification of functional nanoparticles for controlled drug delivery. J Dispers Sci Technol 2003;24:475–87.
[60] Biondi M, Ungaro F, Qualia F. Controlled drug delivery in tissue engineering. Adv Drug Deliv Rev 2008;60:229–42.

[61] Balasundaram G. Nanomaterials for better orthopedics. In: Webster TJ, editor. Nanotechnology for the regeneration of hard and soft tissues. Singapore: World Scientific; 2007. p. 53–78.
[62] Webster TJ, Ejiofor JU. Increased osteoblast adhesion on nanostructured metals: Ti, Ti_6Al_4V, and CoCrMo. Biomaterials 2004;19:4731–9.
[63] Webster TJ, Ergun C, Doremus RH, Siegel RW, Bizios R. Enhanced osteoblast functions on nanophase ceramics. J Biomed Mater Res 2000;51:475–9.
[64] Khang D, Lu J, Yao C, Haberstroh KM, Webster TJ. The role of nanometer and sub-micron surface features on vascular and bone cell adhesion on titanium. Biomaterials 2008;29:970–83.
[65] Tran P, Sarin L, Hurt R, et al. Promising orthopedic materials for bone cancer patient: titanium coated with selenium nanoclusters. In: Proceeding of 34th Northeast Bioengineering Conference, Providence, USA; 2008. p. 262–3.
[66] Kim KH, Kwon TY, Kim SY, et al. Preparation and characterization of anodized titanium surfaces and their effect on osteoblast responses. J Oral Implantol 2006;32:8–13.
[67] Yao C, Webster TJ. Anodization: a promising nano-modification technique of titanium implants for orthopedic applications. J Nanosci Nanotechnol 2006;6:2682–92.
[68] Yao C, Webster TJ. Anodization: a promising nano-modification technique of titanium implants for orthopedic applications. In: Webster TJ, editor. Nanotechnology for the regeneration of hard and soft tissues. Singapore: World Scientific; 2007. p. 79–106.
[69] Yao C, Perla V, McKenzie JL, Slamovich EB, Webster TJ. Anodized Ti and Ti6Al4V possessing nanometer surface features enhances osteoblast adhesion. J Biomed Nanotechnol 2005;66:68–73.
[70] Balasundaram G, Yao C, Webster TJ. TiO_2 nanotubes functionalized with regions of bone morphogenetic protein-2 increases osteoblast adhesion. J Biomed Mater Res A 2007;84A:447–53.
[71] Popat KC, Eltgroth M, LaTempa TJ. Decreased Staphylococcus epidermis adhesion and increased osteoblast functionality on antibiotic-loaded titania nanotubes. Biomaterials 2007;28:4880–8.
[72] Chun AL, Moralez JG, Webster TJ, Fenniri H. Helical rosette nanotubes: a biomimetic coating for orthopedics. Biomaterials 2005;26:7304–9.
[73] Zhang LJ, Chen YP, Rodriguez J, Fenniri H, Webster TJ. Biomimetic helical rosette nanotubes and nanocrystalline hydroxyapatite coatings on titanium for improving orthopedic implants. Int J Nanomedicine 2008;3:323–33.
[74] Zhang L, Rakotondradany F, Myles AJ, Fenniri H, Webster TJ. Arginine-glycine-aspartic acid modified rosette nanotube–hydrogel composites for bone tissue engineering. Biomatcrials 2009;30(7):1309–20.
[75] Zhao B, Hu H, Mandal SK. A bone mimic based on the self-assembly of hydroxyapatite on chemically functionalized single-walled carbon nanotubes. Chem Mater 2005;17:3235–41.
[76] Zanello LP, Zhao B, Hu H. Bone cell proliferation on carbon nanotubes. Nano Lett 2006;6:562–7.
[77] Sirivisoot S, Yao C, Xiao X. Greater osteoblast functions on multiwalled carbon nanotubes grown from anodized nanotubular titanium for orthopedic applications. Nanotechnology 2007;18:365102.
[78] Kim JY, Khang D, Lee JE. Decreased macrophage density on carbon nanotube patterns on polycarbonate urethane. J Biomed Mater Res A 2009;88A:419–26.
[79] Usui Y, Aoki K, Narita N. Carbon Nanotubes with high bone-tissue compatibility and bone-formation acceleration effects. Small 2008;4:240–6.

[80] Kloss FR, Gassner R, Preiner J. The role of oxygen termination of nanocrystalline diamond on immobilization of BMP-2 and subsequent bone formation. Biomaterials 2008;29:2433–42.
[81] Yang L, Sheldon BW, Webster TJ. Orthopedic diamond coatings: control of surface properties and their impact on osteoblast adhesion and proliferation. J Biomed Mater Res A 2009;91A:548–56.
[82] Yang L, Sheldon BW, Webster TJ. The impact of diamond nanocrystallinity on osteoblast functions. Biomaterials 2009;30:3458–65.
[83] Leach JK. Multifunctional cell-instructive materials for tissue regeneration. Regen Med 2006;1:447–55.
[84] Salvay DM, Shea LD. Inductive tissue engineering with protein and DNA-releasing scaffolds. Mol Biosyst 2006;2:36–48.
[85] Ma PX. Biomimetic materials for tissue engineering. Adv Drug Deliv Rev 2008;60:184–98.
[86] Li C, Vepari C, Jin HJ, Kim HJ. Electrospun silk-BMP-2 scaffolds for bone tissue engineering. Biomaterials 2006;27(16):3115–24.
[87] Hosseinkhani H, Hosseinkhani M, Khademhosseini A. Bone regeneration through controlled release of bone morphogenetic protein-2 from 3-D tissue engineered nano-scaffold. J Control Release 2007;117:380–6.
[88] Jabbarzadeh E, Nair LS, Khan YM. Apatite nano-crystalline surface modification of poly (lactide-co-glycolide) sintered microsphere scaffolds for bone tissue engineering: implications for protein adsorption. J Biomater Sci Polym Ed 2007;18:1141–52.
[89] Wei G, Jin Q, Giannobile WV. The enhancement of osteogenesis by nanofibrous scaffolds incorporating rhBMP-7 nanospheres. Biomaterials 2007;28:2087–96.
[90] Piskin E, Isoglu A, Bolgen N, Vargel I. *In vivo* performance of simvastatin-loaded electrospun spiral-wound polycaprolactone scaffolds in reconstruction of cranial bone defects in the rat model. J Biomed Mater Res A 2009;90:1137–51.
[91] Xu HHK, Weir MD, Simon CG. Injectable and strong nanoapatite scaffolds for cell/growth factor delivery and bone regeneration. Dent Mater 2008;24:1212–22.
[92] Yoon JJ, Chung HJ, Lee HJ. Heparin-immobilized biodegradable scaffolds for local and sustained release of angiogenic growth factor. J Biomed Mater Res 2006;79:934–42.
[93] Jeon O, Song SJ, Kang SW. Enhancement of ectopic bone formation by bone morphogenetic protein-2 released from a heparin-conjugated poly (L-lactic-co-glycolic acid) scaffold. Biomaterials 2007;28:2763–71.
[94] Basmanav FB, Kose GT, Hasirci V. Sequential growth factor delivery from complexed microspheres for bone tissue engineering. Biomaterials 2008;29:4195–204.
[95] Wei G, Ma PX. Nanostructured biomaterials for regeneration. Adv Funct Mater 2008;18:3568–82.
[96] Vogelson CT. Advances in drug delivery systems. Modern Drug Discov 2001;4:49–50.
[97] Feynman R. There's plenty of room at the bottom. In: Lecture at an American Physical Society meeting, Caltech, CA. December 29, 1959; 1959.
[98] Taniguchi N. On the basic concept of nano-technology. In: Proc. Intl. Conf. Prod. London, Part II, British Society of Precision Engineering; 1974.
[99] Kayser O, Lemke A, Hernandez-Tejo N. The impact of nanobiotechnology on the development of new drug delivery systems. Curr Pharm Biotechnol 2005;6:3–5.
[100] Thomasin C, Corradin G, Ying M. Tetanus toxoid and synthetic malaria antigen containing poly(lactide)/poly (lactide-co-glycolide) microspheres: importance of polymer degradation and antigen release for immune response. J Control Release 1996;41:131–45.

[101] Sanchez A, Gupta RK, Alonso MJ. Pulsed controlled-released system for potential use in vaccine delivery. J Pharm Sci 1996;85:547–52.
[102] Gupta P, Vermani K, Garg S. Hydrogels: from controlled release to pH-responsive drug delivery. Drug Discov Today 2002;7:569–79.
[103] Kost J, Langer R. Responsive polymeric delivery systems. Adv Drug Deliv Rev 2001;46:125–48.
[104] Miyata T, Uragami T, Nakamae K. Biomolecule-sensitive hydrogels. Adv Drug Deliv Rev 2002;54:79–98.
[105] Berner B, Dinh SM. Electronically controlled drug delivery. London, UK: CRC Press; 1998.
[106] Hafeli UO. Magnetically modulated therapeutic systems. Int J Pharm 2004;277:19–24.
[107] Southwood LL, Frisbie DD, Kawcak CE. Delivery of growth factors using gene therapy to enhance bone healing. Vet Surg 2004;33:565–78.
[108] Rose FR, Hou Q, Oreffo RO. Delivery system for bone growth factors-the new players in skeletal regeneration. J Pharm Pharmacol 2004;56:415–27.
[109] Franceschi RT. Biological approaches to bone regeneration by gene therapy. J Dent Res 2005;84:1093–103.
[110] Rasubala L, Yoshikawa H, Nagata K. Platelet-derived growth factor and bone morphogenetic protein in the healing of mandibular fractures in rats. Br J Oral Maxillofac Surg 2003;41:173–8.
[111] Morishita N, Nakagami H, Morishita R. Magnetic nanoparticles with surface modification enhanced gene delivery of HVJ-E vector. Biochem Biophys Res Commun 2005;334:1121–6.
[112] Rozanova N, Zhang JZ. Metal and magnetic nanostructures for cancer detection, imaging and therapy. J Biomed Nanotechnol 2008;4:377–99.

Frontiers in nanotechnology-enabled orthopedic materials

8

In previous chapters, a large number of nanotechnology-enhanced materials for orthopedics have been covered and discussed by category. These nanomaterials are well known or accepted in the research and translational areas of orthopedics. It is not difficult to reach a conclusion that rapid development of nanoscale fabrication and processing technologies has enabled engineers to incorporate new benefits into orthopedic materials, which would greatly improve the efficacy of implants or clinical outcomes of treatment. Recent advances and progress in this area have been pushing the research and development of orthopedic nanomaterials to a new horizon, featuring by more multidisciplinary, mechanistic, and translational studies. Those trends received substantial attention and interest, although many of these frontier studies are premature and even conceptual. However, these trends represent the future directions of nanotechnology-enabled orthopedic materials and, thus, are worth addressing.

This chapter covers three research area that attract intensive interest from not only materials engineers or physicians but also physicists, cell biologists, and mathematicians. First, multifunctional and smart orthopedic implants that incorporate sensing, imaging, diagnosis, and therapeutic functionalities are discussed. Ideally, these multifunctional and smart platforms allow surgeons or physicians to monitor interactions between biological systems and materials (e.g., immune and blood responses, infection, etc.) and implement necessary intervention or treatment *in situ* and/or automatically. Second, regulation of stem cell fate and functions by nanostructured materials and their potential applications in treating orthopedic degenerative diseases and promoting bone regeneration are analyzed. Finally, recent progress in mechanistic study on how nanomaterials can affect tissue or cell responses in musculoskeletal system is reviewed.

8.1 Smart and multifunctional orthopedic implants

Extraordinary properties of nanomaterials enable the medical implants (including implantable materials and devices) to respond to magnetic, chemical, electrical, temperature, optical, and ultrasound signals *in vivo*, leading to the possibility of developing smart or intelligent orthopedic implants [1]. The concept of smart implants encompasses the ability to respond to biological or environmental changes and ideally the capability to deal with such changes by preventing unwanted consequences or promoting desirable effects. Therefore, with regard to the current orthopedic implant designed for a single purpose or limited functions, smart implants are often multifunctional and expected to serve medical needs better.

Nanotechnology-Enhanced Orthopedic Materials. http://dx.doi.org/10.1016/B978-0-85709-844-3.00008-2

8.1.1 *Total implants with sensing and drug administration abilities*

The sizes of biological macromolecules (e.g., proteins, nucleic acids, and macromolecular assemblies) are comparable to nanoscale building blocks and, thus, nanoscale building blocks can provide a natural bridge between biology and man-made systems. Based on this principle, nanotechnology is extremely promising for instantaneous detection of biomolecules or biosystems (proteins, ligands, pathogens, DNA, etc.) and responding to such entities in real time. Recent advances in nano-/microelectromechanical systems (N/MEMS) indicate that responsive and automated nanoscale sensors, manipulators, and interfaces between human and digital computation may become a reality in the future [2]. For orthopedic application purpose, Webster et al. designed a smart total hip implant capable of biosensing and drug administration with implemented microchips and external devices [3], as illustrated in Figure 8.1. In this system, electrochemical biosensors integrated on the implant surface are designed to determine bone growth, infection, and inflammation around the orthopedic implant; while a responsive coating placed on the biosensors is responsible for the electrically controllable delivery of antibiotic and anti-inflammatory drugs. In addition, the information gathered from the biosensor is expected to be converted into a radio frequency signal and sent to external computer analyzers through a transmitter integrated in the

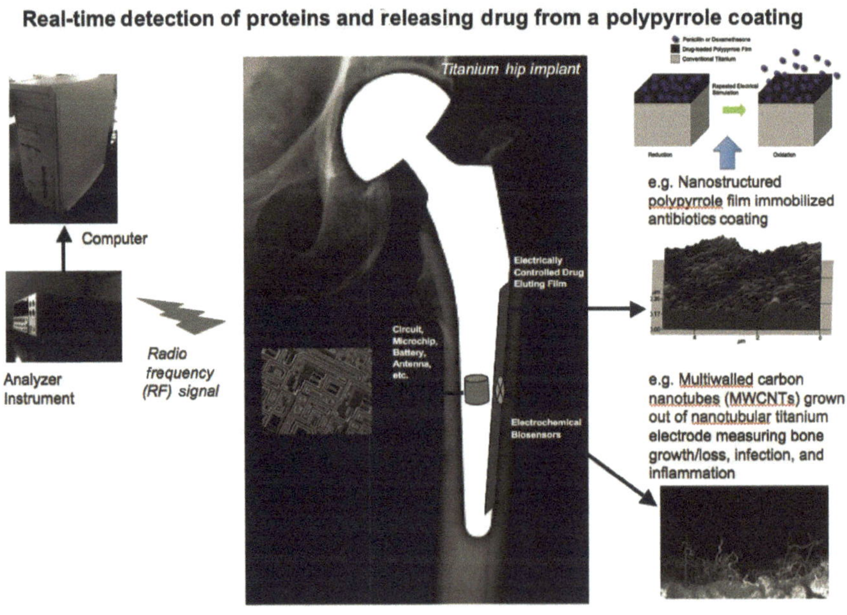

Figure 8.1 Schematic illustrating the concept of a smart orthopedic hip implant with sensing and drug administration abilities. The implant surface is composed of nanotechnology-enabled, electrochemically controlled sensor and drug-eluting film.
Modified with permission of Dr. Webster from Ref. [3].

stem of the hip implant and, in turn, the signal controlling drug delivery in the responsive coating is sent back to the implant after analysis.

Webster et al. has also developed several fundamental nanotechnologies for this system including electrochemical nanosensors and nanofilm for controlled drug delivery, which are both integrated together in the aforementioned smart orthopedic implant. The electrochemical sensors were designed based on multiwalled carbon nanotubes grown out of nanotubular titanium electrode. Electrochemical properties of the nanosensor such as dielectric constant and capacitance are sensitive to molecule or cell attachment and, thus, can be used for monitoring biological events [5]. The biological events include bone growth or loss, bacterial biofilm formation and infection, and acute or chronic inflammatory responses around the implant. Conceptually, information of these biological events is transmitted to external computers for diagnosis and analysis by physician and, if drug intervention is necessary, a signal would be sent to the electrically controlled drug-eluting film to release molecules or therapeutic moieties (Figure 8.1). For the responsive drug delivery, Webster et al. developed a nanostructured polypyrrole film on titanium with the capability to release antibiotics under the repeated reduction–oxidation cycles driven by electrical stimulation [6]. Results from cyclic voltammetry showed that 80% of the model drug (dexamethasone) was released on demand when sweep voltages were applied for five cycles.

Although this nanotechnology-enabled smart hip implant is still under development, it has shown the promise as a self-diagnosing and responding device to detect new bone growth, infection, and other biological events surrounding the implant, allowing doctors to monitor the healing cascade and to diagnose potential infection or inflammation and, ultimately, to intervene in these processes to ensure implant success.

8.1.2 Nanomaterials controlling immune responses, infection, or cancer growth

Future implantable materials should simultaneously enhance tissue regeneration while minimizing immune responses and inhibiting infection, and will be even better if the materials can prevent cancer occurrence [7]. To date, some material formulations with nanotopography or other nanoscale characteristics have been discovered to exhibit multiple capabilities of promoting cell functions, reducing immune responses, inhibiting bacterial and/or cancer growths at the same time. Although there is a lack of mechanistic understanding about why a single material formulation can possess multiple desirable properties for medical implants, the possibility of creating such a "magic material" by nanotechnology has been increased by many recent studies.

Nanotopography of poly(lactic-*co*-glycolic) acid (PLGA) has revealed wanted effects on the responses of tissue, bacteria, and cancer cells. For example, Smith et al. [8] reported that nanorough PLGA surfaces, prepared by NaOH-etching, promoted osteoblast adhesion while inhibiting the adhesion of fibroblast (cell forms connective tissue). Subsequent study further demonstrated that increased osteoblast function but decreased fibroblast adhesion was induced only if the PLGA surface roughness was in the range of 50–100 nm [9]. For bacteria like *Staphylococcus aureus* (a common, dangerous pathogen causing surgical infections), the NaOH-treated PLGA nanorough surfaces revealed significant inhibitory effect without the use of antibiotics. Specifically, a decrease in bacterial growth for up to ~40% after 2 days of culture was achieved on the nanofeatured PLGA membranes compared to untreated controls [10]. More importantly, results also showed that while bacterial growth was stunted, mammalian cell growth was not. For anticarcinogenic properties of PLGA, Zhang and Webster examined the adhesion, proliferation, apoptosis, and vascular endothelial growth factor (VEGF) secretion of breast epithelial adenocarcinoma (BEA) cells and lung epithelial carcinoma (LEC) cells on PLGA nanorough surfaces [11,12]. Results showed that PLGA with 23 nm spherical nanofeatures significantly decreased the functions of BEAs (including decreased proliferation rate, increased apoptosis, and decreased VEGF synthesis) compared to nanosmooth PLGA or PLGA with 300 and 400 nm spherical features (Figure 8.2). Interestingly, while the functions of breast

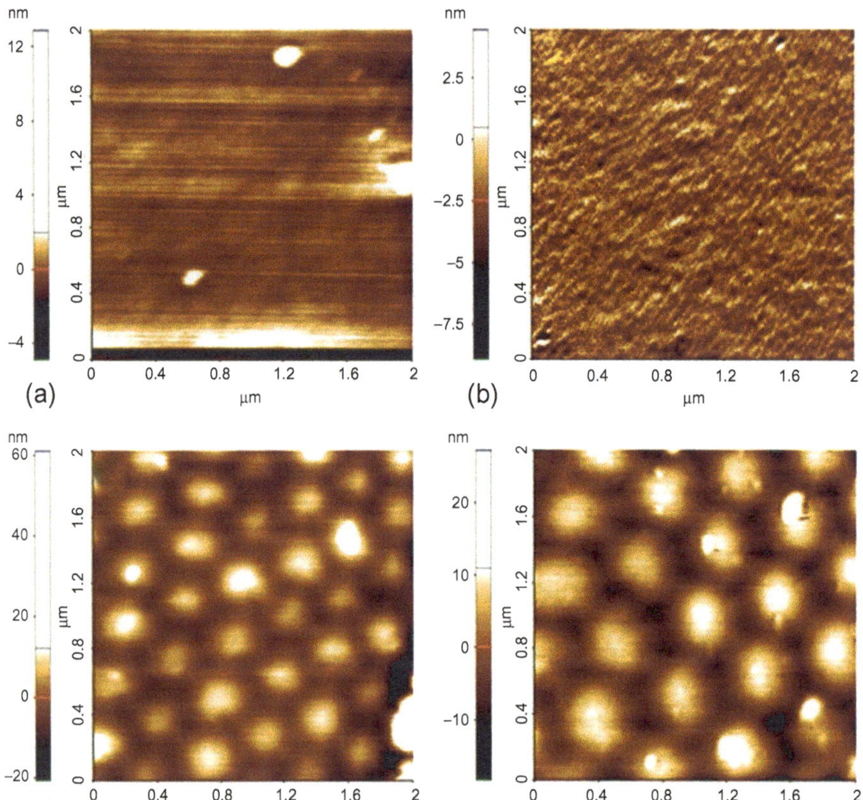

Figure 8.2 Atomic force microscopy (AFM) images of PLGA nanotopographies with spherical features that could selectively alter the adhesion, proliferation, apoptosis, and vascular endothelial growth factor (VEGF) secretion of lung and breast cancer cells: (a) nanosmooth and feature diameters of (b) 23 nm, (c) 300 nm, and (d) 400 nm. See Section 8.1.2. Image reprinted with permission from Ref. [12].

cancer cells were suppressed, the proliferation rate of healthy breast epithelial cells was found to increase by ~24% on the 23 nm featured PLGA surface compared to the nanosmooth ones. For LEC cells and compared to all other PLGA topographies mentioned above, the lowest growth rate was found on the PLGA nanotopography with a feature diameter of 400 nm, while the least VEGF synthesis was observed on the 23 nm featured surface [12].

Nanostructured Ti, well known for its biocompatibility and ability to enhance bone regeneration, is another possible implantable material possessing multiple capabilities including anti-infection property and reduced immune response. For example, Puckett et al. examined *in vitro* adhesion of *Staphylococcus aureus*, *Staphylococcus epidermidis*, and *Pseudomonas aeruginosa* on conventional Ti, nanorough Ti (produced by electron beam evaporation), and nanotubular and nanotextured Ti (produced by two different anodization processes). Results indicated that nanorough Ti surfaces decreased the adhesion of all the bacteria the most compared to conventional Ti (had micron-sized grains) [13]. In addition, anodized Ti with various nanoscale topographies from nanotexture to nanotube, which have been shown to improve bone formation (described in Chapter 2), is reported to possess inhibitory effects on inflammatory cell responses [14]. Specifically, Ti surfaces anodized at 10, 15, and 20 V, corresponding to the surface topographics of nontubular nanotexture, 40–50 nm nanotubes, and 60–70 nm nanotubes, respectively, revealed reduction of macrophage (IC-21 macrophage cell line) adhesion density and cell spreading to different extents compared to conventional Ti. The 60–70 nm nanotubular Ti surfaces showed the most significant effect on decreasing the immune responses of macrophage among all the nanostructured surfaces, indicating an extraordinary opportunity to improve orthopedic implant efficacy by enhancing osseointegration and reducing immune response and infection.

Similarly, nanophased Se also exhibits both anticancer and antibacterial effects in addition to its good biocompatibility properties of supporting or promoting normal bone cell functions (see Chapter 2). Nanophased Se coated on Ti and stainless steel are reported to promote healthy bone tissue growth but prevent carcinogenesis [15,16]. In several *in vitro* studies, noncancerous osteoblasts revealed greater adhesion and proliferation on Se coatings than that on uncoated Ti and stainless steel, indicating that nanophased Se had no adverse effects on healthy osteoblast functions *in vitro*. But more importantly, the function of cancerous osteoblasts on these Se coatings was significantly inhibited compared to the results obtained from non-Se-coated implant materials. Moreover, Se nanoparticles have a strong inhibitory effect against *Staphylococcus aureus* and its biofilm formation [17,18]. For example, an *in vitro* study showed that the selenium nanoparticle coating on polycarbonate could significantly inhibit *Staphylococcus aureus* growth to 8.9% and 27% when compared with an uncoated polycarbonate surface after 24 and 72 h, respectively [17].

Nanotechnology-enabled materials with multiple capabilities of promoting tissue growth and preventing immune response, infection, or carcinogenesis create rich opportunities for the design and manufacture of orthopedic implants with better efficacy and success. Looking for more implantable materials with similar multicapabilities is, therefore, becoming a future trend and direction that is worth further study and development.

8.2 Nanomaterials for the control of stem cell fate and functions

In recent decades, along with the development of stem cell technology and growing interest in utilizing stem cells in regenerative medicine and tissue repair, mediation of stem cell fate by biomaterial property became an important topic. Stem cells are undifferentiated cells that possess remarkable potential to develop into many different cell types and can also divide to produce more stem cells. The multipotency of some kinds of stem cells such as human mesenchymal stem cells (MSCs) is of great importance and promise for the treatment of diseases or medical conditions. In both the clinical and research fields of orthopedics, many therapies and strategies for bone regeneration have been proposed or studied based upon the use of MSCs, scaffolding biomaterials, and/or biomolecules. In the pioneering directions of this field, tailoring material properties to mediate the fate and functions of MSCs has received extraordinary interest from biomaterial engineers. The number of publications in this area has been soaring since the mid-2000s, and significant progress has been achieved in the discovery of material-regulated MSC responses in a variety of materials [19].

For example, Dalby et al. successfully demonstrated the control of MSC differentiation via fabrication of nanoscale symmetry and disorder on polymethylmethacrylate (PMMA) substrates [20]. They created symmetric or disordered nanopits of 120 nm in diameter and 100 nm in depth on the PMMA surfaces, and the cell test results showed that highly ordered or symmetric nanotopographies produced very low or negligible MSC adhesion and osteoblastic differentiation, while MSCs on random nanotopography exhibited a more osteoblastic morphology and underwent more osteospecific differentiation. Another recent study from the same group reported the ability of poly(ε-caprolactone) (PCL) nanotopography to maintain MSC phenotype and multipotency for a long period of time up to 8 weeks [21]. With regard to material composition, nanotopography of both natural and synthetic polymers or macromolecules offers templates for MSC differentiation and self-renewal [19]. Figure 8.3 shows representative natural and synthetic nanopatterns that are successfully fabricated to offer a template for MSC proliferation and differentiation.

Besides topography, the rigidity of nanomaterial can also affect the focal adhesion formation (see Section 8.3.2.2), cytoskeletal organization, and mechanical properties of human MSC [22]. Specifically, MSCs cultured on 350 nm gratings of polystyrene (PS) and poly(dimethylsiloxane) (PDMS) showed decreased expressions of several integrin subunits, compared to that on the nonpatterned substrates, and assumed an elongated morphology accompanied by the alignment of actin cytoskeleton. Furthermore, MSCs on gratings surface of stiffer PS appear to have a lower elastic modulus and apparent viscosity. However, the effects of nanotopography become insignificant when very soft PDMS substrates are used. This result demonstrates that nontopographical factors like material rigidity have a significant impact on stem cell functions and such influence may surpass the effect of nanotopography. Hence, a variety of nanomaterial properties, including topography, rigidity and wettability, must be taken into account when designing high-performance nanomaterials for controlling stem cell fate.

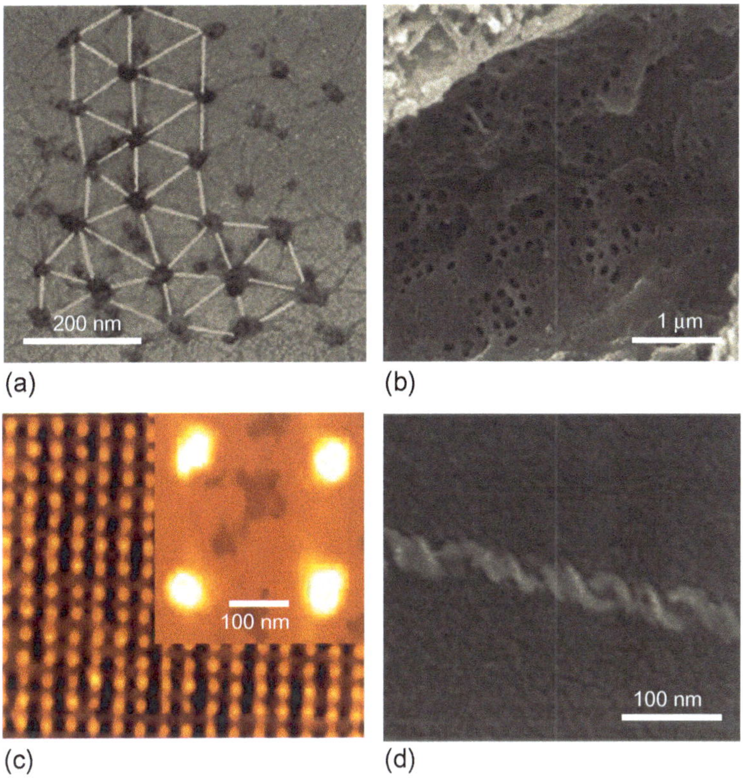

Figure 8.3 Natural (a, b) and synthetic (c, d) nanopatterns that offer templates for MSC differentiation and self-renewal: (a) Type X collagen and (b) fenestrated sinusoidal capillaries, (c) dip-pen patterned nanoislands (70 nm diameter with 28 nm center–center spacing) that retain MSC multipotency, and (d) self-assembled helical amphiphile with 63 nm periodicity that stimulates MSC osteogenesis.
Reprinted with permissions from Ref. [19].

More interestingly, a recent study found that, without osteogenic-inducing media, altering the size of such nanotubular features led to either augmented human MSC adhesion or their direct differentiation into osteoblasts [23]. For instance, small (30 nm in diameter) nanotubular topography promoted MSC adhesion without noticeable differentiation, whereas stem cells on larger (70–100 nm diameter) nanotubular topographies revealed dramatic elongation (10-fold increase), which is believed to induce cytoskeletal stress and subsequently trigger differentiation of MSCs into osteoblast-like cells.

Apparently, the ability of rationally designed nanomaterials to mediate stem cell fate and functions adds a new dimension to stem cell-based therapies and treatments for orthopedic diseases and conditions. The approach of utilizing material physicochemical properties to control stem cell behavior, rather than using biochemical molecules or drugs, has exceptional potential and promise in establishing stem cell therapy with high efficacy but low risk.

8.3 Mechanistic study on nanomaterial-mediated tissue and cell responses

For nanotechnology-enhanced orthopedic materials, there is a compelling need to understand the mechanisms of the interactions between nanomaterials and bone tissues or cells, which would provide important information for designing and fabricating better orthopedic implants or surgical devices. This section overviews recent progress in the experimental and modeling studies on the mechanisms of nanomaterial-mediated tissue or cell responses from two aspects: cell filopodial dynamics on nanostructured materials and related implications, and mechanistic study on the biological events at the cell–materials interface.

8.3.1 Cell filopodial dynamics on nanomaterial

Recently, examining the behavior of individual cells in response to nanotopography has become an intensively studied area, presumably because direct evidence and more insights on cell–nanomaterial interactions can be provided through such studies. Particularly, cell locomotion on a nanomaterial surface has attracted great attention owing to its pivotal role in the processes like wound healing and neural regeneration. For triggering cell locomotion on a biomaterial surface, accumulating evidence has shown that cell filopodia play an important role in the initiation of cell locomotion by sensing and transducing signals (chemical as well as mechanical) from outside into the cell [24]. Filopodium is the finger-like protrusion composed of parallel bundles of actin filaments wrapped around by cell membrane, as illustrated in Figure 8.4. The role of filopodia in serving as sensors for cells to seek microenvironment and sites

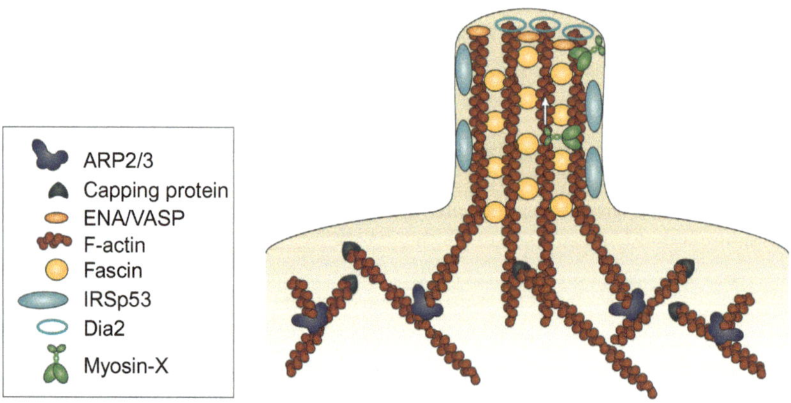

Figure 8.4 Schematic illustrating the molecular structure of filopodium, where bundles of actin filaments (connected by various cross-linking proteins) are wrapped around by the cell membrane.
Modified with permission from Ref. [24].

for adhesion and migration has been well documented [26]. Therefore, understanding filopodial dynamics on biomaterial nanotopography will be critical in analyzing the response of cells to topographical cues.

There are already some studies focusing on the interplay between filopodia and substrates with different rigidities. For example, recent experimental and modeling investigations indicated that filopodia can respond to the stiffness of the microenvironment via a myosin-directed retrograde flow mechanism [27]. Results showed that periodic deformation occured in soft substrates (elastic modulus <1 kPa) due to the slow actin retrograde flow and high traction force generated at the cell–substrate interface. In contrast, little deformation was induced in stiff substrates (with moduli as high as about 100 kPa) because the filopodial traction force on the stiff surface was much smaller. However, this knowledge on filopodium–substrate interactions is clearly inadequate or even inapplicable for most biomaterials whose stiffness is typically in the range of MPa to GPa (such as metals, ceramics, and many polymers). As a result, the traction force generated can hardly deform these stiff materials and a filopodium may have to follow exactly the surface profile during its growth. Therefore, topography of these biomaterials should become a critical factor affecting filopodial dynamics and other related processes such as cell adhesion and spreading.

Following this logic, a number of studies have been performed and found that, indeed, the growth dynamics of filopodia is regulated by the nanotopography of stiff materials. For example, filopodial formation of human osteoblast on microcrystalline diamond (MCD) and nanocrystalline diamond (NCD) substrates, with distinct topographies but similar surface chemistry and hydrophobicity (surface energy), has been examined [28]. Interestingly, filopodia of osteoblasts cultured on MCD films tethered to specific sites on the diamond surface, while radial as well as parallel filopodia were formed on the NCD substrate. It was also reported that fibroblasts growing on well-defined 3-D sharp-tip Si substrate formed only short filopodia along the sharp tips, when the height of nanoposts was around 500–600 nm. In comparison, long filopodia, starting from the tip all the way down to the valley between nanoposts, formed if the height decreased to 50–100 nm (the spacing between nanoposts was fixed at 230 nm) [29]. Another study using the similar Si substrate found that fibroblast filopodia could reach to the bottom surface when the feature height of sharp-tip nanotopographies (nanoposts or nanogrates with a periodicity of 230 nm) was between 50 and 100 nm and, consequently, the cell was well spread. However, when the feature height increases to the range of 200–600 nm, most filopodia failed to reach to the bottom and the cell was unable to conform to the surface nanotopography (instead, only contacted with the sharp tips) [30]. Similarly, Dalby et al. reported that fibroblast filopodia could sense the topography of PMMA substrates, with nanocolumn arrays up to a feature height of ~160 nm, while the topographical sensitivity of filopodia on PCL nano-pit arrays could be as low as 35 nm in pit diameter [31,32].

Based on these preliminary results, important questions on how filopodia interact with stiff topography and how the deformation of a filopodium occurs near the surface features have raised. To answer these questions, computational and modeling approaches become more powerful and effective in providing better visualization and understanding of filopodial dynamics than experimental approaches. In a recent study, osteoblast filopodial extensions and cell spreading were studied through modeling and computational simulations, and simulation results were verified by experiments [28]. A diffusion-deflection model describing the filopodial extension on stiff substrate topography was established, in which a growing filopodium was treated as a rigid beam. The rigid beam clamps at the point from which it protrudes from the cell body and feels attraction from adhesion proteins on the substrate surface (Figure 8.5a). The model-predicted filopodial morphology was in good agreement with the actual morphology of osteoblast filopodial, indicating that the filopodial extension was affected by the substrate topography. Figure 8.5b shows how a substrate topography (blue line) composed of surface feature sizes larger than filopodial diameter could redirect the filopodial extension, resulting in deformed filopodial morphology (curved red line). More importantly, simulations on the nano (surface with random grain sizes 30–80 nm) and submicron (surface with random grain sizes 400–800 nm) topographies for the same time periods (10^5 simulation steps) revealed that the filopodium on nanotopography grew 1.09 times (mean value) faster than that on submicron topography (Figure 8.5c), agreeing with the experimental measurements of osteoblast spreading area. The results suggested that increasing the lateral dimensions or height of nanometer surface features may inhibit cell filopodial extension and subsequently, decrease cell spreading. This study of simulation and experimental verification depicted a possible new approach of modeling and computational simulation to investigate the mechanisms behind nanotopography-mediated filopodial dynamics and pertinent cell functions.

8.3.2 Mechanistic study on cell–nanomaterial interface

The studies summarized in the previous chapters clearly support the idea that nanomaterial plays a crucial role in mediating cell responses/functions (including adhesion, spreading, proliferation, differentiation, and migration). However, the underlying mechanisms as well as what exactly happen at the cell–material interface are still not well understood to date. Undoubtedly, these questions are critical for the rational and effective design and fabrication of biomaterials for orthopedics in the future. Herein, several possible answers that have been proposed and studied are introduced.

8.3.2.1 Selective protein adsorption

There are a great number of studies focusing on whether and how the nanomaterial properties can affect the adsorption of several cell adhesion proteins, especially fibronectin and vitronectin. It is well documented that increase in fibronectin or vitronectin adsorption on material surfaces can promote short-term cell adhesion and, over a prolonged period, proliferation [33,34]. It is also commonly believed that only nanotopography with feature dimensions similar to those of surface-bound proteins (~10 nm) affect their activities [35]. For example, Gonzalez-Garcia et al. studied fibronectin adsorption on well-defined nanotopographies of 14, 29, and 45 nm deep

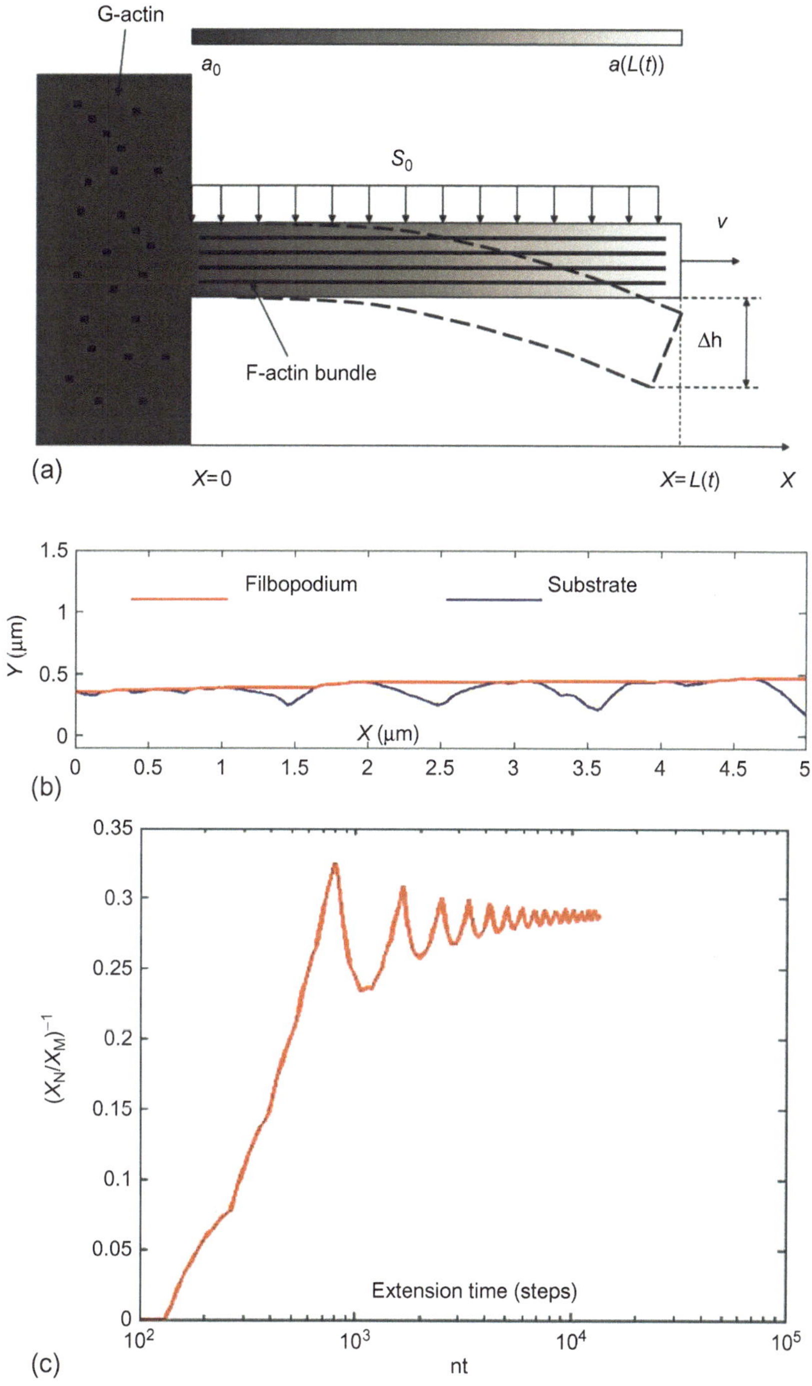

Figure 8.5 (a) Schematic depicting the deflection-diffusion model of filopodial extension; (b) illustration shows the modeling of filopodia (red line) growth on material topography (blue line); (c) simulation-predicted differences in cell filopodia growth distances on a nanotopography (X_N) and a submicron topography (X_M) versus simulation time (expressed by simulation steps).
Modified with permission from Ref. [28].

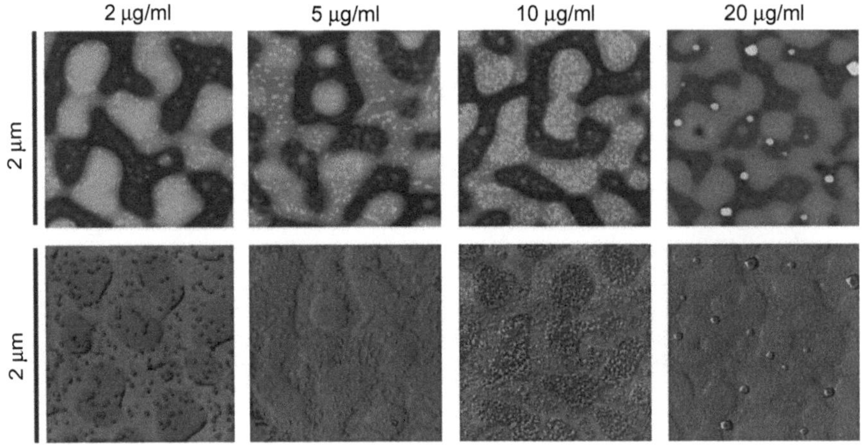

Figure 8.6 Selective adsorption of fibronectin on 14 nm deep-pit nanotopography of PLLA/PS. Height (first row) and phase (second row) AFM images for the nanotopography after fibronectin adsorption from solutions of concentrations 2, 5, 10, and 20 μg/ml. The height image allows one to identify the peaks and valleys in the surface while the phase image provides direct observation of fibronectin distribution.
Reprinted with permission from Ref. [36].

pits on poly(L-lactide) (PLLA)/PS membrane [36]. Fibronectin adsorption quantified by radiolabeling revealed that the amount of protein was higher on the 14 nm deep-pit nanotopography than on the other two surfaces. More interestingly, distribution of fibronectin <10 μg/ml on 14 nm deep-pit topography tends to segregate in valleys (Figure 8.6), while fibronectin of high concentrations or distributed on other nanotopographies appears to be more evenly coated on the surface. These results clearly demonstrate the effect of nanotopography with feature scale around 10 nm on the amount and distribution of fibronectin absorbed on PLLA/PS surface.

To date, however, whether material nanotopography alone is enough to increase the adsorption of specific proteins still remains unclear. Some studies reported a significant increase in fibronectin or vitronectin adsorption on surface topographies in the nanometer regime, which is probably due to the increased surface area or altered hydrophilicity. An investigation found that increasing the roughness of titania surface (from 15 to 30 nm) also led to an increase in adsorbed proteins (including bovine serum albumin, fibrinogen, and streptavidin) and, furthermore, the increase in protein adsorption was actually much higher than that in the surface area. This interesting phenomenon was probably caused by the elevated protein nucleation on the surface, which was promoted by surface nanoscale pores [37]. In contrast, recent studies revealed that the amounts of fibronectin adsorbed per unit true surface area were not statistically different for microcrystalline and NCD films with similar hydrophobicity, but osteoblast and fibroblast adhesion and proliferation were significantly enhanced on the nanocrystalline surface [38,39]. Cai et al. also showed that the nanotopography on Ti had little influence on the adsorption of albumin and fibrinogen [40]. Similarly, a study on the adsorption of hen egg lysozyme on poly(ether sulfone) (PES) surfaces clearly demonstrated that neither the adsorption density nor the structural stability of lysozyme was influenced by variations in the nanometer-scale PES roughness (average roughness values ranging from 5 to 60 nm) [41].

A recent study on ultrahigh molecular weight polyethylene surfaces actually suggest that nanotopography, chemistry, crystallinity, and molecular chain anisotropy could affect protein and assembly at the same time [42]. It is possible, as demonstrated by the study, that nanotopography loses its effectiveness on mediating protein adsorption when surface chemistry and hydrophobicity are changed. Therefore, although some studies point toward the position that selective adsorption of adhesion proteins on material nanotopography might be the key for its influence on cell functions [34,43,44], more investigations are needed to uncover how nanotopography interacts with other material properties and collectively influences protein adsorption and assembly [33]. Unfortunately, existing experimental techniques still have limitations in probing real-time, ultrasmall-scale phenomena such as protein behaviors and, hence, may not be sufficient to answer questions of this kind. To this end, modeling and computational approaches may play a much important role in understanding and predicting protein adsorption on nanomaterials in the near future.

8.3.2.2 Focal adhesion

Another way to probe the underlying mechanisms of nanomaterial-mediated cell responses utilizes a molecular biology approach and focuses on the so-called focal adhesions, by which close contact between the cell and extracellular matrix (ECM) is established. The formation of focal adhesion requires the assembly of various proteins, which then physically link the cell cytoskeleton (composed of F-actin) to the ECM, as illustrated in Figure 8.7. Focal adhesions also serve as signaling hubs where information from outside (like mechanical or chemical stimuli) can be transmitted

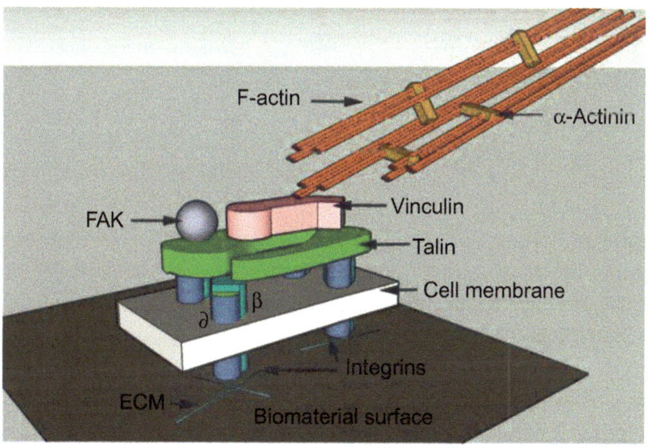

Figure 8.7 Illustration of the molecular structure of focal adhesion.
Reprinted with permission from Ref. [45].

into the cell and, subsequently, affect its behavior. As such, it is conceivable that focal adhesions may play a key role in the observed nanomaterial-mediated cell responses. Many studies have been conducted along this line and readers may refer to a thorough review by Biggs et al. for more details [45].

The formation of focal adhesion is a highly complex process. In general, specific binding between α- and β-chain transmembrane proteins (i.e., integrins) and corresponding motifs in ECM molecules (e.g., the RGD motif in fibronectin and vitronectin) occurs first. Integrins are then laterally reinforced by the assembly of various adhesion proteins like vinculin and talin (Figure 8.7). With increased integrin recruitment, large and matured focal adhesion plaques will emerge that can then be connected to the cytoskeleton/cell nucleus through F-actin bundles (often referred to as stress fibers) [45]. Many studies have been conducted to examine how material nanotopography influences focal adhesion-related kinases, including focal adhesion kinase (FAK), which regulates integrin-dependent signaling pathways, and extracellular signal-regulated kinases (ERK) 1 and 2 that are activated by FAK and mediate cell differentiation and survival. For example, it has been shown that the enhanced integrin–material interaction and cell spreading, induced by nanotopography, can up-regulate the expressions of FAK and ERK1/ERK2 in osteoprogenitor cells [46,47]. Other studies reported that nanoscale features can influence or even disrupt focal adhesion formation through the perturbation of integrin activation and clustering. For example, by restricting possible cell–substrate adhesion to discrete functionalized dots where only one integrin molecule can attach, Arnold et al. [48] revealed, for the first time, that there was a critical spacing (~70 nm) between individual dots for stable cell–ECM adhesion. It has been found that tight adhesion patches can be formed if dots are spaced more closely together than this distance. On the other hand, no apparent adhesion was observed for spacings larger than this critical value. Evidently, although there are lots of unknowns, existing evidence supports that nanomaterial properties have significant impact on the formation and functioning of focal adhesions and, hence, could possibly regulate cell behavior through integrin-specific signaling pathways.

8.4 Summary and future perspectives

The three sections in this chapter—smart and multifunctional implants, nanomaterials controlling stem cell responses, and mechanistic study on the nanomaterial-mediated cell responses—represent some new frontiers in the research of nanotechnology-enhanced orthopedic materials. Clearly, the frontier research in this field encompasses a wide spectrum of experimental and theoretic approaches from biology to engineering to physics. It is not difficult to foresee that future studies on nanotechnology-enabled biomaterials would point toward the materials with high efficacy and low risk, smart and responsive capabilities, and better understanding of the mechanisms behind the interaction between such materials and biological systems. Therefore, there are several future trends in these directions that are of particular interest and importance.

First, better smart and multifunctional orthopedic nanomaterials with more capabilities are expected. In the near future, nanomaterial platforms are expected that integrate diagnostic, imaging or monitoring, therapeutic, and tissue regenerative abilities and can be minimal-invasively deployed or administrated. Such nanomaterial systems would complete several tasks that are currently performed individually by different materials or devices with better efficacy and safety. In addition, as implied by the example of a smart hip implant in Section 8.1.1, rapid development of N/MEMS may also contribute to the emergence of novel orthopedic implants based on such novel nanomaterials.

Second, functional nanomaterials that can guide stem cell differentiation and regulate its functions will continue to be an important research area. In the past few years, a large number of *in vivo* studies have supported the safety and efficacy of nanomaterial-enhanced stem cell therapy for possible human applications, including treating bone diseases like osteoarthritis and vertebral disk degeneration [49]. The functional nanomaterials have strongly demonstrated their important roles in interacting and influencing stem cell fate or functions toward the wanted directions and, therefore, should be further studied and expedited for possible clinical applications. As a result, a soaring number of translational studies on applying nanomaterials and stem cells to treat clinical conditions are expected in the next decade.

Regarding the mechanistic studies on the interaction between nanomaterial and tissue or cell, so far the mechanisms behind the interactions still remain largely unclear. On one hand, although the influence of nanomaterials is ubiquitously observed in many types of tissue cells, there is not a general trend on how nanomaterial properties could modulate cell or tissue behaviors. On the other hand, direct and *in situ* observation of subcellular structures and biomolecules that play key roles in cell–material interactions, are still very difficult or even impossible. To this end, development of new techniques and interdisciplinary approaches including computational and modeling approaches may open new avenues for addressing these critical challenges.

On the aspect of experimental study, developing nanofabrication, biodetection, and imaging techniques will receive more attention in the near future because they allow us to closely monitor various events at the cell–material interface, including direct and *in situ* observation of the fate and behavior of key biomolecules and subcellular structures like filopodium and cell membrane. On the modeling and simulation side, one possible trend is studying the influence of nanomaterials on cell adhesion by focusing on both specific or nonspecific protein interactions. Also, the study on filopodial dynamics is expected to reach a new level and how to take into account the 3-D nature of the problem and understanding the retrograde flow or the molecular motor-driven active transport of actin would be elucidated. As a result, the research outcomes from both the experimental and theoretical studies on the interactions between nanomaterials and biological systems could definitely provide new insights into the design, fabrication, and applications of better nanomaterials for orthopedics.

References

[1] Yang L, Zhang LJ, Webster TJ. Nanobiomaterials: state of the art and future trends. Adv Eng Mater 2011;13:B197–217.

[2] Yang L, Webster T. Monitoring tissue healing through nanosensors. In: Webster TJ, editor. Nanotechnology enabled in situ sensors for monitoring health. New York: Springer; 2011. p. 41–59.

[3] Webster TJ. Novel medical devices based on nanotechnologies: the future is here. In: ATA world biomedical device congress. Shanghai; 2014.

[4] Deleted in proof.

[5] Sirivisoot S, Webster TJ. In situ bone growth detection using carbon nanotubes–titanium sensors. BioNanoScience 2013;3:184–91.

[6] Sirivisoot S, Pareta R, Webster TJ. Electrically controlled drug release from nanostructured polypyrrole coated on titanium. Nanotechnology 2011;22:085101.

[7] Khang D, Carpenter J, Chun YW, Pareta R, Webster TJ. Nanotechnology for regenerative medicine. Biomed Microdevices 2010;12:575–87.

[8] Smith LL, Niziolek PJ, Haberstroh KM, Nauman EA, Webster TJ. Decreased fibroblast and increased osteoblast adhesion on nanostructured NaOH-etched PLGA scaffolds. Int J Nanomed 2007;2:383.

[9] Vance RJ, Miller DC, Thapa A, Haberstroh KM, Webster TJ. Decreased fibroblast cell density on chemically degraded poly-lactic-*co*-glycolic acid, polyurethane, and polycaprolactone. Biomaterials 2004;25:2095–103.

[10] Karahaliloglu Z, Ercan B, Chung S, Taylor E, Denkbas EB, Webster TJ. Nanostructured anti-bacterial poly-lactic-*co*-glycolic acid films for skin tissue engineering applications. J Biomed Mater Res A 2014;102:4598–608.

[11] Zhang L, Webster TJ. Poly-lactic-glycolic-acid surface nanotopographies selectively decrease breast adenocarcinoma cell functions. Nanotechnology 2012;23:155101.

[12] Zhang L, Webster TJ. Decreased lung carcinoma cell functions on select polymer nanometer surface features. J Biomed Mater Res A 2012;100:94–102.

[13] Puckett SD, Taylor E, Raimondo T, Webster TJ. The relationship between the nanostructure of titanium surfaces and bacterial attachment. Biomaterials 2010;31:706–13.

[14] Rajyalakshmi A, Ercan B, Balasubramanian K, Webster TJ. Reduced adhesion of macrophages on anodized titanium with select nanotube surface features. Int J Nanomed 2011;6:1765.

[15] Tran P, Webster TJ. Enhanced osteoblast adhesion on nanostructured selenium compacts for anti-cancer orthopedic applications. Int J Nanomed 2008;3:391.

[16] Tran PA, Sarin L, Hurt RH, Webster TJ. Titanium surfaces with adherent selenium nanoclusters as a novel anticancer orthopedic material. J Biomed Mater Res A 2010;93:1417–28.

[17] Wang Q, Webster TJ. Nanostructured selenium for preventing biofilm formation on polycarbonate medical devices. J Biomed Mater Res A 2012;100:3205–10.

[18] Tran PA, Webster TJ. Selenium nanoparticles inhibit *Staphylococcus aureus* growth. Int J Nanomed 2011;6:1553.

[19] Dalby MJ, Gadegaard N, Oreffo RO. Harnessing nanotopography and integrin–matrix interactions to influence stem cell fate. Nat Mater 2014;13:558–69.

[20] Dalby MJ, Gadegaard N, Tare R, Andar A, Riehle MO, Herzyk P, et al. The control of human mesenchymal cell differentiation using nanoscale symmetry and disorder. Nat Mater 2007;6:997–1003.

[21] McMurray RJ, Gadegaard N, Tsimbouri PM, Burgess KV, McNamara LE, Tare R, et al. Nanoscale surfaces for the long-term maintenance of mesenchymal stem cell phenotype and multipotency. Nat Mater 2011;10:637–44.

[22] Yim EK, Darling EM, Kulangara K, Guilak F, Leong KW. Nanotopography-induced changes in focal adhesions, cytoskeletal organization, and mechanical properties of human mesenchymal stem cells. Biomaterials 2010;31:1299–306.
[23] Oh S, Brammer KS, Li YJ, Teng D, Engler AJ, Chien S, et al. Stem cell fate dictated solely by altered nanotube dimension. Proc Natl Acad Sci 2009;106:2130–5.
[24] Mattila PK, Lappalainen P. Filopodia: molecular architecture and cellular functions. Nat Rev Mol Cell Biol 2008;9:446–54.
[25] Deleted in proof.
[26] Zhang H, Berg JS, Li Z, Wang Y, Lang P, Sousa AD, et al. Myosin-X provides a motor-based link between integrins and the cytoskeleton. Nat Cell Biol 2004;6:523–31.
[27] Chan CE, Odde DJ. Traction dynamics of filopodia on compliant substrates. Science 2008;322:1687–91.
[28] Yang L, Chinthapenta V, Li Q, Stout D, Liang A, Sheldon BW, et al. Understanding osteoblast responses to stiff nanotopographies through experiments and computational simulations. J Biomed Mater Res A 2011;97A:375–82.
[29] Choi CH, Hagvall SH, Wu BM, Dunn JC, Beygui RE, Kim C-J. Cell interaction with three-dimensional sharp-tip nanotopography. Biomaterials 2007;28:1672–9.
[30] Choi C-H, Heydarkhan-Hagvall S, Wu BM, Dunn JC, Beygui RE, Kim C-J. Cell growth as a sheet on three-dimensional sharp-tip nanostructures. J Biomed Mater Res A 2009;89:804–17.
[31] Dalby MJ, Riehle MO, Sutherland DS, Agheli H, Curtis ASG. Changes in fibroblast morphology in response to nano-columns produced by colloidal lithography. Biomaterials 2004;25:5415–22.
[32] Dalby MJ, Gadegaard N, Riehle MO, Wilkinson CDW, Curtis ASG. Investigating filopodia sensing using arrays of defined nano-pits down to 35 nm diameter in size. Int J Biochem Cell B 2004;36:2005–15.
[33] Wilson CJ, Clegg RE, Leavesley DI, Pearcy MJ. Mediation of biomaterial–cell interactions by adsorbed proteins: a review. Tissue Eng 2005;11:1–18.
[34] Webster TJ, Ergun C, Doremus RH, Siegel RW, Bizios R. Specific proteins mediate enhanced osteoblast adhesion on nanophase ceramics. J Biomed Mater Res 2000;51:475–83.
[35] Kasemo B, Gold J. Implant surfaces and interface processes. Adv Dent Res 1999;13:8–20.
[36] González-García C, Sousa SR, Moratal D, Rico P, Salmerón-Sánchez M. Effect of nanoscale topography on fibronectin adsorption, focal adhesion size and matrix organisation. Colloids Surf B: Biointerfaces 2010;77:181–90.
[37] Scopelliti PE, Borgonovo A, Indrieri M, Giorgetti L, Bongiorno G, Carbone R, et al. The effect of surface nanometre-scale morphology on protein adsorption. PLoS One 2010;5:e11862.
[38] Yang L, Li YW, Sheldon BW, Webster TJ. Altering surface energy of nanocrystalline diamond to control osteoblast responses. J Mater Chem 2012;22:205–14.
[39] Yang L, Zhang L, Webster TJ. Carbon nanostructures for orthopedic medical applications. Nanomedicine-UK 2011;6:1231–44.
[40] Cai K, Frant M, Bossert J, Hildebrand G, Liefeith K, Jandt KD. Surface functionalized titanium thin films: zeta-potential, protein adsorption and cell proliferation. Colloids Surf B: Biointerfaces 2006;50:1–8.
[41] Roach P, Farrar D, Perry CC. Surface tailoring for controlled protein adsorption: effect of topography at the nanometer scale and chemistry. J Am Chem Soc 2006;128:3939–45.
[42] Keller TF, SchoÃànfelder J, Reichert J, Tuccitto N, Licciardello A, Messina GM, et al. How the surface nanostructure of polyethylene affects protein assembly and orientation. ACS Nano 2011;5:3120–31.

[43] Khang D, Kim SY, Liu-Snyder P, Palmore GT, Durbin SM, Webster TJ. Enhanced fibronectin adsorption on carbon nanotube/poly(carbonate) urethane: independent role of surface nano-roughness and associated surface energy. Biomaterials 2007;28:4756–68.

[44] Miller DC, Haberstroh KM, Webster TJ. PLGA nanometer surface features manipulate fibronectin interactions for improved vascular cell adhesion. J Biomed Mater Res A 2007;81:678–84.

[45] Biggs MJP, Richards RG, Dalby MJ. Nanotopographical modification: a regulator of cellular function through focal adhesions. Nanomed Nanotechnol Biol Med 2010;6:619–33.

[46] Biggs MJ, Richards RG, Gadegaard N, McMurray RJ, Affrossman S, Wilkinson CD, et al. Interactions with nanoscale topography: adhesion quantification and signal transduction in cells of osteogenic and multipotent lineage. J Biomed Mater Res A 2009;91:195–208.

[47] Sjostrom T, Dalby MJ, Hart A, Tare R, Oreffo RO, Su B. Fabrication of pillar-like titania nanostructures on titanium and their interactions with human skeletal stem cells. Acta Biomater 2009;5:1433–41.

[48] Arnold M, Cavalcanti-Adam EA, Glass R, Blummel J, Eck W, Kantlehner M, et al. Activation of integrin function by nanopatterned adhesive interfaces. ChemPhysChem 2004;5:383–8.

[49] Yang L, Sheldon BW, Webster TJ. Nanophase ceramics for improved drug delivery: current opportunities and challenges. Am Ceram Soc Bull 2010;89:24–31.

Safety of nanotechnology-enhanced orthopedic materials

9

As increasing numbers of nanomaterials have been studied, fabricated and used in research and even consumer products such as paints, textiles and sunscreens, safety and toxicity issues related to nanomaterials have become important concerns. Until now, both *in vitro* and *in vivo* evidence has suggested that these issues may have great impacts on people's daily life and the development of nanotechnology. In general, safety issues are associated with almost every aspect of nanomaterials, including research and development, fabrication and modification, transportation and disposal, and certainly application. Unfortunately, the biological responses to and toxicity of nanomaterials have not been sufficiently understood, partially due to the complexity of the problem and the lack of well-established evaluation methods. Each step moving toward the understanding of nanomaterial safety is challenging because this problem spans across different length scales (from atom sizes to the scales of atmospheric motion) and a variety of disciplines (from materials science to ecology), adding exceptional uncertainty and difficulty to research and study.

In medical fields, safety is absolutely the first priority for any nanotechnology-related researches and applications. Due to the size of nanomaterials or nanodevices—small enough to penetrate cells and tissues unnoticed—their toxicity becomes an extremely important concern when used for long-term applications inside the human body. It is, therefore, necessary and mandatory to investigate the health impacts of nanomaterials before ultimately using and manufacturing these materials for human-related purposes. With regard to orthopedic materials, the investigation on nanomaterial safety needs to be considered from as least two aspects: the nanomaterials composing orthopedic implant and the post-implantation nanoscale wear debris or degradation products from implants. In this chapter, emphasis is given to the former aspect, while some information on the safety of wear or degradation debris is also provided.

9.1 Overview of host responses to biomaterials

At tissue and cellular levels, safety and toxicity of implantable biomaterials are closely related to a series of complicated physiological activities triggered at the biomaterial-tissue interface. These physiological activities are usually referred to as host responses taking place at a time scale from milliseconds to months. Figure 9.1 lists the host responses that often occur toward biomaterials when implanted *in vivo*. Several important host responses are briefly reviewed below.

Nanotechnology-Enhanced Orthopedic Materials. http://dx.doi.org/10.1016/B978-0-85709-844-3.00009-4

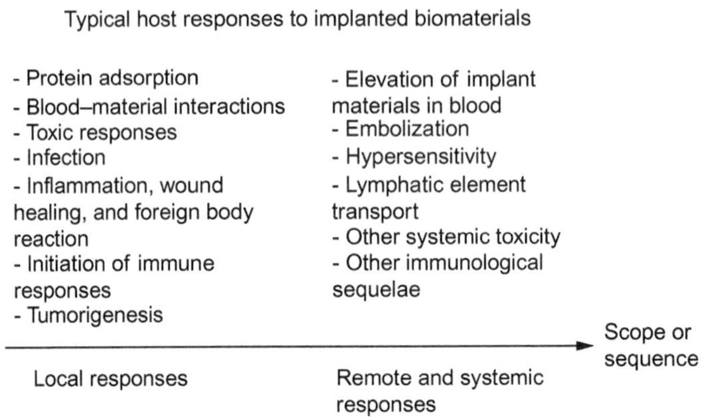

Figure 9.1 Typical host responses to implanted biomaterials.
Reprinted with permission from Ref. [1].

9.1.1 Inflammatory reactions

In a broad sense, inflammatory reactions comprise a sequence of events such as acute and chronic inflammations, granulation tissue generation, foreign body reaction, and fibrous encapsulation (or fibrosis). Depending on the implantation process and material properties, the inflammatory responses to the biomaterials do not necessarily include all these sequential events. The progression of inflammatory reactions is mediated by a number of chemical mediators released from plasma, cells, and tissues, including cytokines, growth factors, platelet activating factors, plasma proteases, vasoactive amines, lysosomal proteases and so on. Acute inflammation is one of the most common inflammatory responses to implanted biomaterials. Acute inflammation starts with the exudation of fluid and plasma proteins and the migration of neutrophils or other defensive cells. Subsequently, activation of neutrophils and macrophages lead to phagocytosis of foreign materials, which may last till the stage of chronic inflammation. Macrophage plays an important role in the inflammation by secreting a number of chemical mediators (e.g., neutral proteases, oxygen free radicals, and growth factors) that regulate inflammatory responses and possibly cause systemic effects in tissues or organs. Granulation tissue formation is a process in which fibroblasts and vascular endothelial cells proliferate and start to generate granulation tissue near the implanted biomaterial. While in the foreign body reaction, which may last for as long as the lifetime of the biomaterial, macrophages, fibroblasts, and/or foreign body giant cells attempt to engulf and phagocytose the biomaterials. Lastly, fibrosis usually takes place at the end-stage of inflammatory responses and is featured by the formation of a fibrous capsule on biomaterials by connective tissue.

9.1.2 Immune responses

Implanted biomaterials are usually recognized as "foreigners" by immune system and inevitably trigger immunological responses that protect host tissue from exposure to such "foreigners." In other words, biomaterials may be treated as a kind of pathogens by the complex network of the immune system including humoral and cellular components. In brief, humoral and cell-mediated immune responses lead to the activation of host defenses and production of protective antibodies and antigen processing cells, which attempt to remove or isolate the foreign materials. Immune responses, however, are possible to cause excessive defending responses that may potentially damage host tissues. Therefore, control or suppression of immune responses to biomaterials is of great interest in biomaterial design. Due to the complexity of immune responses and space limitations, the pathology associated with immune responses to biomaterials is not covered here and readers can refer to the literature (e.g., Ref. [2]).

9.1.3 Toxicity

Toxicity of biomaterials is usually viewed at both cellular and systemic levels. Toxicity at the cellular level, also known as cytotoxicity, is often caused by direct chemical toxicity of biomaterials, inflammatory reactions, or immune responses to biomaterials. Excessive or severe cytotoxicity, inflammatory reactions, or immune responses may lead to the toxicity remote from the initial insult and affecting organs or organ systems, which is defined as systemic toxicity. It is important to emphasize that toxicity caused by biomaterials is usually dose dependent. For cytotoxicity or nonimmune systemic toxicity, a threshold below which the biomaterial reveals little toxicity may be carefully determined by *in vitro* and *in vivo* studies. However, for immune systemic toxicity, determination of such a threshold is extremely difficult because immune responses are individual-dependent and largely affected by properties, dosage, and implantation location of the biomaterials. Toxicity of nanomaterials will be further discussed in this chapter.

9.1.4 Thrombosis

Blood–material interaction is apparently important for implant materials that are in contact with blood in surgery or service. Hemostatic mechanisms provide prevention and control of bleeding by formation of a clot, and this process can be greatly affected by the presence of biomaterials. In general, the hemostatic reactions include primary and secondary hemostasis. Shortly after a biomaterial is in contact with blood, the primary hemostasis initiates a sequence of biological events. First, coagulation proteins and platelets attach onto the material surface and then start releasing granule contents. Second, the granule contents stimulate more platelets to migrate and aggregate on the material surface, forming a platelet plug on the material surfaces. At the same time, secondary hemostasis kicks in to form a fibrin gel that strengthens the platelet plug (platelet thrombus). Because excessive thrombi coagulation may cause severe adverse consequences, control of thrombosis on biomaterial surface is of great importance.

9.1.5 Infection

Implant-associated infection is one of the major reasons behind implant failure, amputation, or even death of patients. Statistics between 2005 and 2006 in the United States reveal that 14.8% of revised total hip arthroplasties were due to infection [3].

Two common bacteria, *Staphylococcus epidermidis* existing on the human skin and *Staphylococcus aureus* that is a natural tissue pathogen, are responsible for most implant-associated infections. Infection often refers to a complex process associated with bacterial adhesion onto material surfaces, aggregation or dispersion of bacterial colonies, and subsequent biofilm formation. Biofilm is a thick biocomposite consisting of a large number of interconnected bacteria, extracellular substances, and other environmental adsorbents. Biofilms on biomaterial surface are responsible for a wide variety of implant-associated infections. However, eradication of biofilm becomes very difficult once it forms, because it develops strong resistance to antimicrobial agents [4]. Moreover, subtherapeutic levels of β-lactam antibiotics have recently been found to induce *Staphylococcus aureus* biofilm formation [5]. Interestingly, degradation products from biomaterials may also trigger implant-associated infection, because they can provide surfaces or niches for bacterial adhesion as well as nutrients (e.g., Fe^{3+}) and stimuli (e.g., Ca^{2+}) for bacterial adhesion, growth, and propagation [1].

9.2 Toxicological effects of the nanomaterials for orthopedic applications

9.2.1 *Metallic nanomaterials*

Owing to prevalent use in orthopedic research and applications, toxicological effects of metallic nanomaterials have been widely studied both *in vitro* and *in vivo*. So far, most of studies have focused on synthetic nanoparticles of Fe Ni, Ti, and Ti alloys, or nanoparticles degraded from metallic implants. Results from these toxicity studies reveal that, in addition to dosage, toxicological effects of metallic nanoparticles are strongly correlated to particle size and surface modification. For example, Ti and TiO_2 particles could be internalized by human neutrophils and cause soft tissue inflammation when they are less than 3 μm [6]. Other studies indicated that, when nanoparticle size decreases below 10–50 nm, the body's defensive responses (e.g., immune response) to nanoparticles may reduce but the likelihood of nanoparticles reaching target organs through the circulatory system is increased [6,7]. Sarmiento-Gonzalez et al. studied Ti levels in the organs and blood of rats with Ti wires implanted into femurs for 18 months. Results reveal that, in the absence of wear, Ti levels in the blood or other organs (such as the heart, lung, spleen, liver, kidney, etc.) were higher than the basal Ti level measured in the rats without Ti implants. This study suggests that nanoscale corrosion or degradation products of the Ti implant can cause systemic accumulation of Ti in the body.

Nanoscale debris from CoCrMo, which is also a widely used orthopedic implant material, may also induce DNA and chromosome damage as well as cytotoxicity. For instance, CoCr nanoparticles demonstrated more severe DNA damage, chromosomal damage, and toxic effects compared to micron-sized CoCr particles [8,9]. For orthopedic implants made of stainless steels, Fe or Ni nanoparticles are also possible sources for triggering toxicity and adverse effects at local or systemic levels. For example, Ni particles implanted in rat soft tissues were found to cause just allergic reactions when their sizes were in the micron regime, while inducing carcinogenicity when the sizes decreased to 500 nm [6]. Nanosized Fe and iron oxide particles as well as Fe^{3+} ions were tested in gills and blood cells of blue mussels [10]. Results showed that such nanoparticles and irons decreased lysosomal stability in blood cells and increased lipid peroxidation in the gills, revealing significant toxicological effects against tissue or organ.

Owing to great potential for medical imaging and drug delivery applications, noble metallic nanoparticles have attracted lots of attention and their toxicological effects have also been studied. Many studies agree that bare noble metallic nanoparticles were bioinert or biocompatible even after they entered cytoplasm. Tkachenko et al. [11] found that Au nanoparticles were readily internalized into the cytoplasm but did not enter the nucleus; therefore, cell viability maintained greater than 95% after 12 h of incubation. Shukla et al. [12] tested the effect of Au nanoparticles on the immunological response of cells by examining the proliferation, nitric oxide, and reactive oxygen species (ROS) production of macrophage cells. After 48 h of up to 100 μM of Au nanoparticles treatment, macrophage cells showed greater than 90% viability with no increase in proinflammatory cytokines TNF-a and IL-1b. However, there are also studies revealing that the toxicological effects of Au nanoparticles depend on cell lines and particle surface properties. For example, Tkachenko et al. [13] reported varied cell viability to the same peptide-BSA-gold nanoparticles in HeLa and 3T3/NIH cell lines. Goodman et al. [14] reported similar cytotoxicity of the same gold nanoparticles among COS-1 cells, red blood cells, and *Escherichia coli*, but with a more toxic effect of ammonium-modified, cationic Au nanoparticles than carboxylate-modified, anionic gold nanoparticles. Another recent study by Chompoosor et al. [15] reported that acute cytotoxicity and genotoxicity of positively charged Au nanoparticles depend on the hydrophobicity of the ligands attached. They showed that increasing hydrophobicity of Au nanoparticles increased their cytotoxicity and ROS production, though DNA damage was decreased. However, conflicting cytotoxicity and DNA damage results for similar Au nanoparticles were reported by others [16], indicating the complexity and difficulty in the toxicological evaluation of metallic nanomaterials.

9.2.2 Polymeric nanomaterials

Emergence of new polymeric nanomaterials has opened a new avenue for the design and fabrication of novel prosthetic implants and tissue engineering scaffolds for treating orthopedic problems. As introduced in Chapter 4, a large number of nanopolymers, polymer-based nanocomposites, and nanostructured hydrogels have demonstrated their potential for creating next-generation orthopedic implants or scaffolds with improved efficacy and clinical outcomes. At the same time, safety and toxicology evaluations on these nanomaterials have been conducted in cell and animal models but only a very few have entered the stage of human trials. Among the toxicology studies to date, most of the nanopolymers and nanocomposites have reported acceptable biocompatibility or low toxicity *in vitro* and *in vivo*. For example, one of the well-accepted nontoxic formulations is poly(ethylene glycol) (PEG), which possesses reduced immunogenicity and little toxic effect. A large variety of nanoparticles that are coated or modified by PEG have not shown PEG-specific toxicity or immune responses [17,18]. Similarly, poly(hydroxyethyl methacrylate) (PHEMA) and poly(*N*-isopropylacrylamide) (PNIPAAm) hydrogels are also reported to be highly biocompatible and nontoxic in most of the studies [19]. Due to space limitations, similar results on other polymeric formulations are not summarized here.

However, even if polymers or composites themselves are not toxic, polymeric materials are more likely to wear, degrade, or decompose *in vivo*, resulting in debris or irritating substances that may induce protein and cell damages or inflammatory reactions to host tissue. Furthermore, this likelihood for biodegradable nanomaterials (e.g., poly(lactic acid) (PLA), poly(glycolic acid) (PGA), poly(lactic-co-glycolic acid) (PLGA)) may drastically increase because of their substantially higher surface areas compared to conventional micron-sized materials [20].

For polymeric nanomaterials, another source of toxicity comes from unreacted monomers, residual reaction agents (e.g., cross-linking initiators), or additives (e.g., surfactants) in the fabrication processes [21]. These substances may leach out and release to tissue when the nanopolymers or nanocomposites interact with body fluids or degrade. For instance, monomers of *N*-isopropylacrylamide and methyl methacrylate that forms PNIPAAm and poly(methyl methacrylate) (PMMA), respectively, are toxic and carcinogenic or teratogenic. As a result, medical products based on PNIPAAm and PMMA always require careful control of fabrication parameters and processes to minimize the unreacted monomers or residues [22].

9.2.3 Ceramic nanomaterials

In general, bioactive and bioresorbable ceramic nanoparticles like calcium phosphates or bioglass are considered to possess good biocompatibility properties and not cause (or cause negligible) adverse tissue responses, but more verification and evaluation are absolutely needed before further human applications of these nanomaterials. For other ceramics related to orthopedic applications, such TiO_2, Fe_2O_3 or Fe_3O_4, Al_2O_3, Cr_2O_3, SiO_2, and ZnO, mixed or even contradictory results in both *in vitro* and *in vivo* studies have been reported. These contradictions indeed require a must know on the toxicity before future applications of ceramic nanoparticles.

For Al_2O_3 nanoparticles, most studies to date conclude that they are biocompatible and reveal little or no adverse effects. This conclusion has been verified in various tissue cells, zebrafish embryo, and fruit flies [23,24]. For example, a recent dose-dependent study on human fibroblasts showed that Al_2O_3 nanoparticles (at doses from 0.1 to 10 mg/T75 flask) had no adverse effect on cell viability and DNA up to 5 days [24]. However, there are a few cases that reported the opposite results. For example, Oesterling et al. [25] showed that the existence of Al_2O_3 nanoparticles increased the cytokine release and adhesion of activated monocytes, indicating proinflammatory responses induced by the nanoparticles.

It is reported that TiO_2 (anatase) nanoparticles with crystallite size ~5 nm could insert into the DNA double helix of mouse liver cell and cause DNA cleavage as well as subsequent cell damage [26]. An early cell study also showed that TiO_2 nanoparticles could produce free radicals with a strong oxidizing ability that can cause DNA damage [27]. However, *in vitro* and *in vivo* studies by Landsiedel et al. [28] revealed that TiO_2 nanoparticles did not show genotoxicity, contradicting the results mentioned above. An *in vivo* study using zebrafish as an animal model also concluded that TiO_2 nanoparticles did not reveal developmental toxic effects on zebrafish embryos and larvae [23].

A great number of opposing results have been reported for the cytotoxicity of both bare and surface-modified iron oxide nanoparticles, and the well-known debate on whether iron oxide nanoparticles are biocompatible or toxic at a variety of doses still remains [29]. Similarly, many studies support that SiO_2 nanoparticles with a size range from the nanometer to submicron are not toxic and SiO_2 has been used as a nontoxic, negative control in cytotoxicity studies of other nanomaterials [30]. But there are also studies showing that SiO_2 nanoparticles of 21 and 48 nm can induce oxidative stress, cell damage, and eventually apoptosis in hepatic and myocardial cells [31,32].

The inconsistency in the aforementioned studies on a variety of ceramic nanoparticles reflects complexity and difficulty in determining and understanding nanomaterial toxicity. There are at least three possible reasons behind this inconsistency. First, toxicity of nanoparticles is strongly dose-dependent and determining a unified threshold of toxicity is difficult for nanomaterials varying in chemistry and microstructures [31–34]. Second, toxicity of nanomaterial may be tissue- or cell-specific, leading to different tissue or cell responses to the same nanomaterial [30]. Third, agglomeration and dissolution in the nanoparticles due to interactions with body fluids (or cell culture media), proteins, or cellular components can add substantial complexity to the experimental results [29]. Inevitably, due to such complexity and difficulty, more investigation and new research methods are indeed necessary to explore nanotoxicity.

9.2.4 Carbon nanostructures

Chapter 5 reviews a number of carbon nanostructures with particular interest to orthopedic applications, including fullerene (C_{60}), carbon nanotubes (CNTs) and carbon nanofibers (CNFs), graphene, and nanodiamonds. In the past few decades, health risks and toxicological data of these newly emerging carbon nanomaterials have been studied but not yet completely known.

Dose and cell-dependent toxicity has been identified with C_{60} and surface modification of C_{60} also has significant effects on the toxicity. A number of studies support that C_{60} has low cellular uptake, no apoptotic effect to tissue cells, and no toxicity to macrophages at concentrations less than 60 μg/mL [29,35,36]. But Sayes et al. reported the dose-dependent cytotoxicity of pristine and surface-modified C_{60} in human dermal fibroblasts and liver carcinoma cells at concentrations from 0.24 to 2400 ppb [37]. They also showed that the surface-modified C_{60} (water-soluble hydroxylated C_{60}) was less toxic than pristine C_{60}. However, C_{60} modified with amino acid revealed dose-dependent proinflammatory responses and toxicity in human epidermal keratinocytes (HEK) at concentrations from 40 to 400 μg/mL [38]. Based on these mixed results, it is difficult if not impossible to draw a conclusion on the safety of C_{60}, and thorough evaluation is needed depending on the specific application.

Based on the published data to date, CNTs appear to have more consistent results of cytotoxicity. For example, human embryo kidney cells, human keratinocytes, HeLa cells, and lung carcinoma cells all exhibited dose- and size-dependent (or aspect ratio-dependent, in some studies) apoptosis and retarded proliferation when exposed to single-walled CNT (SWCNT) [39,40]. This dose- and time-dependent toxicity has also been observed with multiwalled CNT (MWCNT) [41]. Besides cytotoxicity, MWCNT may also induce immune and inflammatory responses (reflected by overexpression of related genes) at high concentrations above 0.6 μg/mL [42]. The dose- and size-dependent toxicity with SWCNTs and MWCNTs is also affected by surface modification. For instance, water-soluble SWCNTs and the SWCNTs coated by carboxyl, biotin, and fluorescein demonstrated reduced cytotoxicity compared to pristine SWCNTs [43,44]. For MWCNTs, impact of surface modifications on cytotoxicity has also been investigated. Results from lung tumor cell tests suggest that MWCNTs functionalized through an acid treatment were more toxic than unmodified MWCNTs at concentrations from 0.002 to 0.2 μg/mL [45].

For nanodiamonds, there are relatively less studies on their toxicity compared to the aforementioned carbon nanostructures. However, results from the limited studies consistently suggest that nanodiamonds are bioinert and biocompatible to a variety of cell types. A number of recent studies all showed low toxicity of nanodiamonds to osteoblasts, fibroblasts, human kidney cells, neuroblastoma, macrophage, keratinocyte, and PC-12 cells, as well as little production of ROS [46,47]. Direct transmission electron microscopy (TEM) evidence also showed that nanodiamond particles were internalized by neuroblastoma cells but did not cause apparent toxicological effect to cell mitochondrial functions [47].

9.3 Mechanisms behind the toxicological effects of nanomaterials

9.3.1 ROS generation

From a biochemistry perspective, material-induced toxicity and inflammation are closely associated with ROS including superoxide anions, superoxide dismutase, hydrogen peroxide and so on. ROS are normal products of cellular metabolism but abnormally high level of ROS can cause harmful effects on cells or tissues, causing damage of DNA, denaturation of proteins, disruption of lipids, or deactivation of specific enzymes [48]. These harmful effects may cause further adverse responses or damages at tissue or organ levels, eventually leading to systemic inflammation or toxicity. Increased ROS level has been widely accepted as a molecular mechanism of nanomaterial-induced toxicity [49,50]. This mechanism is clearly elucidated by the results from ZnO nanoparticles shown in Figure 9.2, where a strong correlation between ROS level and cytotoxicity (quantified as the effective concentration corresponding to 50% cell viability, EC50) is demonstrated [50].

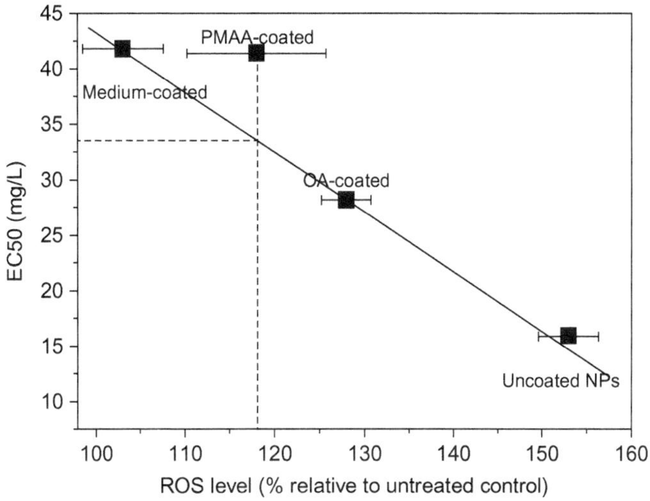

Figure 9.2 Relationship between levels of ROS and cytotoxicity of WIL2-NS cells (as the effective concentration corresponding to 50% cell viability, EC50) for the various ZnO nanoparticles.
Reprinted with permission from Ref. [50].

A variety of studies on Fe, ZnO, SiO_2, CNT and C_{60} nanoparticles have confirmed the generations of excessive free radicals, ROS, and oxidative stress when they expose to cells, causing protein oxidation, lipid peroxidation, DNA damage, cell membrane or cytoskeleton disruption, and eventually, apoptosis or cell injury [10,29,32,37,50–52]. In fact, ROS generation is ubiquitously associated with the interaction between nanomaterials and biological systems [53], as illustrated in Figure 9.3. In the figure, mechanisms by which nanomaterials interact with biological systems are considered from four aspects: material composition, electronic structure, bonded surface spices (e.g., coating metals and organics), and solubility. For each aspect, based on limited toxicity data, generation of ROS plays a role in the mechanism that explains how nanomaterial properties can lead to toxicity. For example, electron capture in the interaction of electron donor or acceptor active sites with molecular dioxygen (O_2) can lead to the formation of the superoxide radical (O_2^-), which through dismutation can generate additional ROS (Figure 9.3) [53]. Also, nanoparticles containing transition metals (e.g., Fe and V) and also coated with redox-cycling organic chemicals (e.g., quinones) can also facilitate a similar process of transferring dioxygen to ROS.

There are two things about ROS generation need to be emphasized. First, although ROS generation is currently the best-developed paradigm for explaining nanotoxicity, detailed mechanisms of ROS generation still remain unclear. Second, despite the toxicity data to date supporting that high level of ROS generation and oxidative stress could lead to nanomaterial toxicity, it is still not clear whether these experimental findings can be directly transferred to evaluate clinical toxicity.

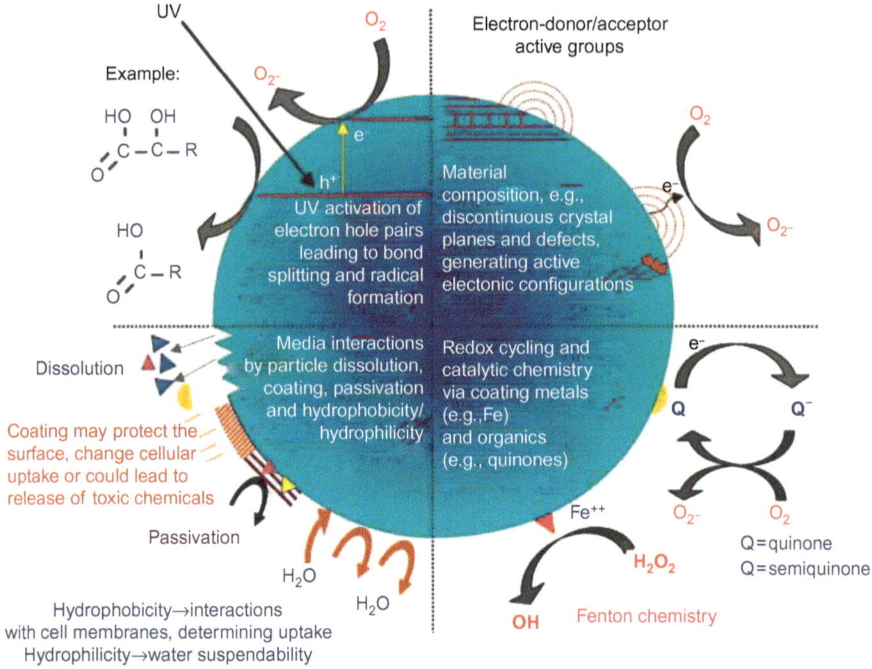

Figure 9.3 The role of ROS in the possible mechanisms by which nanomaterials interact with biological systems. ROS generation is associated with all the four aspects of the mechanisms, in which examples illustrate the importance of material composition, electronic structure, bonded surface species (e.g., metal-containing), surface coatings (active or passive), and solubility, including the contribution of surface species and coatings and interactions with other environmental factors (e.g., UV activation).
Reprinted with permission from Ref. [53].

9.3.2 *Cellular uptake of nanomaterials*

Cellular uptake is an important process by which cells internalize substances such as nutrients, proteins and foreign materials, and therefore plays an important role in cell–nanomaterial interactions. The endocytic pathways for the internalization of substance include phagocytosis, macropinocytosis, clathrin-mediated endocytosis, and non-clathrin-mediated endocytosis [54,55], all of which are closely related to the host responses addressed in the previous section. However, cellular uptake of a material does not necessarily implicate toxicity. But due to exceptionally small size and large surface area, cellular uptake of nanomaterials is different from that of conventional micron materials, which probably induce altered toxicological effects and, thus, needs to be studied.

Most of the studies to date have investigated the endocytosis of nanomaterials with sizes less than 200 nm [54,55]. Results from these studies clearly suggest that cellular uptake of nanomaterials is size-dependent. On one hand, the type of endocytic pathways for the internalizing nanomaterials is highly dependent on material size. For example, latex nanospheres of less than 200 nm are readily endocytosed by nonphagocytic cells via clathrin-mediated endocytosis, while the endocytosis pathway switches to caveolae internalization (i.e., non-clathrin-mediated) when the sphere size increased to 500 nm [56]. On the other hand, endocytic efficiency is also a function of the material size. For instance, latex nanoparticles with sizes of 50–100 nm can be internalized rapidly by cells, whereas the cellular uptake of nanoparticles with sizes of 200 nm or more is substantially slower [56].

Surface physicochemical properties of nanomaterial are also crucial factors strongly affecting the cellular uptake process. Usually, the surface properties refer to surface area and roughness, topography, surface charge, surface chemistry, and surface wettability (or surface energy). Under *in vitro* or *in vivo* conditions, these surface properties can lead to aggregation or agglomeration, self-assembled structures or superstructures, and dissolution, which may also affect cellular uptake. Surface properties of nanomaterial interact with endocytic receptors on cell membranes and, therefore, are able to regulate cellular internalization processes of nanomaterials. For example, cellular uptake of CNTs is found to be strongly dependent on their physicochemical properties. Recently, a unified explanation to how CNTs are internalized depending on their size and surface properties was summarized by Raffa et al. [57], as shown in Figure 9.4. For micron-sized aggregates of CNTs formed through surface charge or surface hydrophobic interactions, or the nonagglomerated CNTs with lengths more than 1 μm, the internalization pathway is phagocytosis. However, for more hydrophobic CNTs, or when CNTs form supramolecular complexes via binding proteins and other ligands, the receptor-mediated endocytosis pathway dominates. For CNTs whose surface properties prevent the formation of supramolecular complexes, the diffusion pathway becomes possible.

Although a large number of studies have investigated the cellular uptakes of many nanomaterials, the understanding of this field is still not sufficient. Fundamental mechanisms underlying internalization pathways and intracellular traffic at the molecular level are not well understood. More importantly, the correlation between cellular uptake and nanomaterial toxicity remains unclear. Further exploration of these issues will be necessary for the design and fabrication of nanotechnology-enhanced orthopedic materials.

9.4 Summary

Growing interest in the exploration of nanomaterials for human-related applications has placed the safety of nanomaterials into an extremely important position. Thorough understanding of nanomaterial safety is arguably a prerequisite for any nanotechnology-related application, but medical applications of any nanomaterials would demand such understanding be as clear as possible. Unfortunately, there are increasing concerns on safety and toxicity issues related to nanomaterials in the past few years, yet not much has been clearly understood so far. Recently, substantial progress has been made in uncovering cytotoxicity of nanomaterials *in vitro* and its related mechanisms such as ROS generation and cellular internalization, but *in vivo* evaluations with well-accepted conclusions are still scarce. Even among the limited

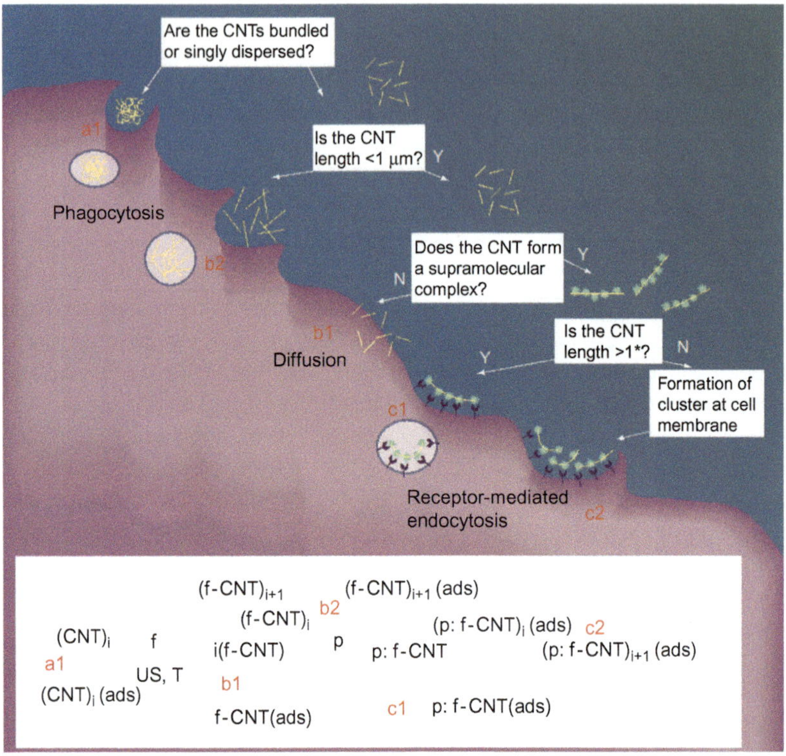

Figure 9.4 Surface properties and size influencing the pathways of CNT internalization. Inserted panel: Nanotube bundles $(CNT)_i$, functionalized with (macro)molecules (f) with the aid of sonication (US) and/or temperature (T), are dispersed in single nanotubes f-CNT; this f-CNT can lead to the formation of supramolecular structure p: f-CNT. (a1) $(CNT)_i$ interact with cell membrane and are internalized via phagocytosis; (b1) f-CNT enter the cells by diffusion; (b2) nanotube dispersion is not stable and nanotubes cluster in solution (nanotube clusters enter the cell by phagocytosis); (c1) p: f-CNT enters the cell by ligand-receptor (l) binding; (c2) p: f-CNT cluster on cell membrane and enter by receptor-mediated endocytosis. CNT: Carbon nanotube; f-CNT: Functionalized CNT.
Reprinted with permission from Ref. [57].

data about nanomaterial-induced toxicity and adverse effects, contradictory results are often presented. As a result, the influence of nanotechnology-enhanced materials on general human health and specific medical applications like orthopedic implants is still not well understood.

Before manufacturing and using nanomaterials for human applications in orthopedics, it is necessary to further investigate the health and safety impacts of these materials from at least three perspectives. First, the relationships between toxicity and the size, dosage, and surface properties of nanomaterial need to be clarified and quantified if possible. Second, better understanding of the pathways and mechanisms by which nanomaterials cause cell damage or adverse tissue responses is wanted. Third, toxic responses to nanomaterials generated from the degradation of implants, wear debris of artificial joints, and residue from materials processing and manufacture, should not be neglected and needs to be clarified [1]. Undoubtedly, timely and systematic efforts on studying these aspects will benefit and facilitate the development of nanotechnology-enhanced orthopedic implants and devices.

References

[1] Yang L, Webster T. Biological responses to and toxicity of nanoscale implant materials. In: Eliaz N, editor. Degradation of implant materials. New York: Springer; 2013. p. 481–508.
[2] Anderson JM, Gristina AG, Hanson SR, Harker LA, Johnson RJ, Merritt K, et al. Host reactions to biomaterials and their evaluation. In: Ratner BD, Hoffman AS, Schoen FJ, Lemons JE, editors. Biomaterials science: an introduction to materials in medicine. San Diego: Academic Press; 1996. p. 165–213.
[3] Bozic KJ. The increasing number of THA revisions in the United States: why is it happening? Orthop Today Eur 2009;29:6.
[4] Mah T-FC, O'Toole GA. Mechanisms of biofilm resistance to antimicrobial agents. Trends Microbiol 2001;9:34–9.
[5] Kaplan JB, Izano EA, Gopal P, Karwacki MT, Kim S, Bose JL, et al. Low levels of β-lactam antibiotics induce extracellular DNA release and biofilm formation in Staphylococcus aureus. MBio 2012;3:e00198-12.
[6] Watari F, Abe S, Koyama C, Yokoyama A, Akasaka T, Uo M, et al. Behavior of in vitro, in vivo and internal motion of micro/nano particles of titanium, titanium oxides and others. J Ceram Soc Jpn 2008;116:1–5.
[7] Vinogradov SV, Bronich TK, Kabanov AV. Nanosized cationic hydrogels for drug delivery: preparation, properties and interactions with cells. Adv Drug Deliv Rev 2002;54:135–47.
[8] Papageorgiou I, Yin ZR, Ladon D, Baird D, Lewis AC, Sood A, et al. Genotoxic effects of particles of surgical cobalt chrome alloy on human cells of different age in vitro. Mutat Res Fundam Mol Mech Mutagen 2007;619:45–58.
[9] Papageorgiou I, Brown C, Schins R, Singh S, Newson R, Davis S, et al. The effect of nano- and micron-sized particles of cobalt-chromium alloy on human fibroblasts in vitro. Biomaterials 2007;28:2946–58.
[10] Kadar E, Lowe DM, Sole M, Fisher AS, Jha AN, Readman JW, et al. Uptake and biological responses to nano-Fe versus soluble FeCl3 in excised mussel gills. Anal Bioanal Chem 2010;396:657–66.
[11] Tkachenko AG, Xie H, Coleman D, Glomm W, Ryan J, Anderson MF, et al. Multifunctional gold nanoparticle-peptide complexes for nuclear targeting. J Am Chem Soc 2003;125:4700–1.
[12] Shukla R, Bansal V, Chaudhary M, Basu A, Bhonde RR, Sastry M. Biocompatibility of gold nanoparticles and their endocytotic fate inside the cellular compartment: a microscopic overview. Langmuir 2005;21:10644–54.
[13] Tkachenko AG, Xie H, Liu YL, Coleman D, Ryan J, Glomm WR, et al. Cellular trajectories of peptide-modified gold particle complexes: comparison of nuclear localization signals and peptide transduction domains. Bioconjug Chem 2004;15:482–90.
[14] Goodman CM, McCusker CD, Yilmaz T, Rotello VM. Toxicity of gold nanoparticles functionalized with cationic and anionic side chains. Bioconjug Chem 2004;15:897–900.

[15] Chompoosor A, Saha K, Ghosh PS, Macarthy DJ, Miranda OR, Zhu ZÄ, et al. The role of surface functionality on acute cytotoxicity, ROS generation and DNA damage by cationic gold nanoparticles. Small 2010;6:2246–9.

[16] Aguirre JD, Angeles-Boza AM, Chouai A, Pellois J-P, Turro C, Dunbar KR. Live cell cytotoxicity studies: documentation of the interactions of antitumor active dirhodium compounds with nuclear DNA. J Am Chem Soc 2009;131:11353–60.

[17] Zahr AS, Davis CA, Pishko MV. Macrophage uptake of core-shell nanoparticles surface modified with poly(ethylene glycol). Langmuir 2006;22:8178–85.

[18] Wang AZ, Gu FX, Farokhzad OC. Nanoparticles for cancer diagnosis and therapy. Safety of nanoparticles. New York: Springer; 2009. p. 1–27.

[19] Meenach SA, Anderson KW, Hilt JZ. Hydrogel nanocomposites: biomedical applications, biocompatibility, and toxicity analysis. Safety of nanoparticles. New York: Springer; 2009. p. 1–27.

[20] Wei X, Lee Y-k, Huh KM, Kim S, Park K. Safety and efficacy of nano/micro materials. Safety of nanoparticles. New York: Springer; 2009. p. 1–26.

[21] Peppas NA, Bures P, Leobandung W, Ichikawa H. Hydrogels in pharmaceutical formulations. Eur J Pharm Biopharm 2000;50:27–46.

[22] Weng H, Zhou J, Tang LP, Hu ZB. Tissue responses to thermally-responsive hydrogel nanoparticles. J Biomater Sci Polym Ed 2004;15:1167–80.

[23] Zhu XS, Zhu L, Duan ZH, Qi RQ, Li Y, Lang YP. Comparative toxicity of several metal oxide nanoparticle aqueous suspensions to Zebrafish (Danio rerio) early developmental stage. J Environ Sci Health A Tox Hazard Subst Environ Eng 2008;43:278–84.

[24] Tsaousi A, Jones E, Case CP. The in vitro genotoxicity of orthopaedic ceramic (Al2O3) and metal (CoCr alloy) particles. Mutat Res 2010;697:1–9.

[25] Oesterling E, Chopra N, Gavalas V, Arzuaga X, Lim EJ, Sultana R, et al. Alumina nanoparticles induce expression of endothelial cell adhesion molecules. Toxicol Lett 2008;178:160–6.

[26] Li N, Ma LL, Wang J, Zheng L, Liu J, Duan YM, et al. Interaction between nano-anatase TiO2 and liver DNA from mice in vivo. Nanoscale Res Lett 2010;5:108–15.

[27] Dunford R, Salinaro A, Cai LZ, Serpone N, Horikoshi S, Hidaka H, et al. Chemical oxidation and DNA damage catalysed by inorganic sunscreen ingredients. FEBS Lett 1997;418:87–90.

[28] Landsiedel R, Ma-Hock L, Van Ravenzwaay B, Schulz M, Wiench K, Champ S, et al. Gene toxicity studies on titanium dioxide and zinc oxide nanomaterials used for UV-protection in cosmetic formulations. Nanotoxicology 2010;4:364–81.

[29] Lewinski N, Colvin V, Drezek R. Cytotoxicity of nanoparticles. Small 2008;4:26–49.

[30] Brunner TJ, Wick P, Manser P, Spohn P, Grass RN, Limbach LK, et al. In vitro cytotoxicity of oxide nanoparticles: comparison to asbestos, silica, and the effect of particle solubility. Environ Sci Technol 2006;40:4374–81.

[31] Ye YY, Liu JW, Chen MC, Sun LJ, Lan MB. In vitro toxicity of silica nanoparticles in myocardial cells. Environ Toxicol Pharmacol 2010;29:131–7.

[32] Ye Y, Liu J, Xu J, Sun L, Chen M, Lan M. Nano-SiO2 induces apoptosis via activation of p53 and Bax mediated by oxidative stress in human hepatic cell line. Toxicol In Vitro 2010;24:751–8.

[33] Gupta AK, Gupta M. Cytotoxicity suppression and cellular uptake enhancement of surface modified magnetic nanoparticles. Biomaterials 2005;26:1565–73.

[34] Jeng HA, Swanson J. Toxicity of metal oxide nanoparticles in mammalian cells. J Environ Sci Health A Tox Hazard Subst Environ Eng 2006;41:2699–711.

[35] Fiorito S, Serafino A, Andreola F, Bernier P. Effects of fullerenes and single-wall carbon nanotubes on murine and human macrophages. Carbon 2006;44:1100–5.

[36] Jia G, Wang HF, Yan L, Wang X, Pei RJ, Yan T, et al. Cytotoxicity of carbon nanomaterials: single-wall nanotube, multi-wall nanotube, and fullerene. Environ Sci Technol 2005;39:1378–83.
[37] Sayes CM, Fortner JD, Guo W, Lyon D, Boyd AM, Ausman KD, et al. The differential cytotoxicity of water-soluble fullerenes. Nano Lett 2004;4:1881–7.
[38] Rouse JG, Yang JZ, Barron AR, Monteiro-Riviere NA. Fullerene-based amino acid nanoparticle interactions with human epidermal keratinocytes. Toxicol in Vitro 2006;20:1313–20.
[39] Cui D, Tian F, Ozkan CS, Wang M, Gao H. Effect of single wall carbon nanotubes on human HEK293 cells. Toxicol Lett 2005;155:73–85.
[40] Manna SK, Sarkar S, Barr J, Wise K, Barrera EV, Jejelowo O, et al. Single-walled carbon nanotube induces oxidative stress and activates nuclear transcription factor-kappa B in human keratinocytes. Nano Lett 2005;5:1676–84.
[41] Monteiro-Riviere NA, Inman AO. Challenges for assessing carbon nanomaterial toxicity to the skin. Carbon 2006;44:1070–8.
[42] Ding LH, Stilwell J, Zhang TT, Elboudwarej O, Jiang HJ, Selegue JP, et al. Molecular characterization of the cytotoxic mechanism of multiwall carbon nanotubes and nano-onions on human skin fibroblast. Nano Lett 2005;5:2448–64.
[43] Kam NWS, Jessop TC, Wender PA, Dai HJ. Nanotube molecular transporters: internalization of carbon nanotube-protein conjugates into mammalian cells. J Am Chem Soc 2004;126:6850–1.
[44] Sayes CM, Liang F, Hudson JL, Mendez J, Guo WH, Beach JM, et al. Functionalization density dependence of single-walled carbon nanotubes cytotoxicity in vitro. Toxicol Lett 2006;161:135–42.
[45] Magrez A, Kasas S, Salicio V, Pasquier N, Seo JW, Celio M, et al. Cellular toxicity of carbon-based nanomaterials. Nano Lett 2006;6:1121–5.
[46] Yu SJ, Kang MW, Chang HC, Chen KM, Yu YC. Bright fluorescent nanodiamonds: no photobleaching and low cytotoxicity. J Am Chem Soc 2005;127:17604–5.
[47] Schrand AM, Huang HJ, Carlson C, Schlager JJ, Osawa E, Hussain SM, et al. Are diamond nanoparticles cytotoxic? J Phys Chem B 2007;111:2–7.
[48] Brooker RJ. Genetics: analysis & principles. New York: Granite Hill Publishers; 2009.
[49] Xia T, Kovochich M, Nel A. The role of reactive oxygen species and oxidative stress in mediating particulate matter injury. Clin Occup Environ Med 2005;5:817–36.
[50] Yin H, Casey PS, McCall MJ, Fenech M. Effects of surface chemistry on cytotoxicity, genotoxicity, and the generation of reactive oxygen species induced by ZnO nanoparticles. Langmuir 2010;26:15399–408.
[51] Sayes CM, Gobin AM, Ausman KD, Mendez J, West JL, Colvin VL. Nano-C-60 cytotoxicity is due to lipid peroxidation. Biomaterials 2005;26:7587–95.
[52] Lin WS, Xu Y, Huang CC, Ma YF, Shannon KB, Chen DR, et al. Toxicity of nano- and micro-sized ZnO particles in human lung epithelial cells. J Nanopart Res 2009;11:25–39.
[53] Nel A, Xia T, Mädler L, Li N. Toxic potential of materials at the nanolevel. Science 2006;311:622–7.
[54] Lanone S, Boczkowski J. Biomedical applications and potential health risks of nanomaterials: molecular mechanisms. Curr Mol Med 2006;6:651–63.
[55] Goldberg M, Langer R, Jia X. Nanostructured materials for applications in drug delivery and tissue engineering. J Biomater Sci Polym Ed 2007;18:241–68.
[56] Rejman J, Oberle V, Zuhorn IS, Hoekstra D. Size-dependent internalization of particles via the pathways of clathrin-and caveolae-mediated endocytosis. Biochem J 2004;377:159–69.
[57] Raffa V, Ciofani G, Vittorio O, Riggio C, Cuschieri A. Physicochemical properties affecting cellular uptake of carbon nanotubes. Nanomedicine 2010;5:89–97.

Index

Note: Page numbers followed by *f* indicate figures and *t* indicate tables.

FSC
www.fsc.org
MIX
Papier | Fördert
gute Waldnutzung
FSC® C083411

Kolibri 360 GmbH
Ferdinand-Jühlke-Straße 7
99095 Erfurt, Deutschland
produktsicherheit@kolibri360.de